Electrochemical Energy Storage Devices and Supercapacitors

An overview

IOP Series in Renewable and Sustainable Power

The IOP Series in Renewable and Sustainable Power aims to bring together topics relating to renewable energy, from generation to transmission, storage, integration, and use patterns, with a particular focus on systems-level and interdisciplinary discussions. It is intended to provide a state-of-the-art resource for all researchers involved in the power conversation.

Series Editor
Professor David Elliott
Open University, UK

About the Editor
David Elliott is emeritus Professor of Technology Policy at the Open University, where he developed courses and research on technological innovation, focusing on renewable energy policy. Since retirement, he has continued to write extensively on that topic, including a series of books for IOP Publishing and a weekly blog post for *Physics World* (physicsworld.com/author/david-elliott)

About the Series
Renewable and sustainable energy systems offer the potential for long-term solutions to the world's growing energy needs, operating at a broad array of scales and technology levels. The IOP Series in Renewable and Sustainable Power aims to bring together topics relating to renewable energy, from generation to transmission, storage, integration, and use patterns, with a particular focus on systems-level and interdisciplinary discussions. It is intended to provide a state-of-the-art resource for all researchers involved in the power conversation.

We welcome proposals in all areas of renewable energy including (but not limited to): wind power, wave power, tidal power, hydroelectric power, PV/solar power, geothermal power, bioenergy, heating, grid balancing and integration, energy storage, energy efficiency, carbon capture, fuel cells, power to gas, electric/green transport, and energy saving and efficiency.

Authors are encouraged to take advantage of electronic publication through the use of colour, animations, video, data files, and interactive elements, all of which provide opportunities to enhance the reader experience.

Do you have an idea for a book you'd like to explore?
We are currently commissioning for the series; if you are interested in writing or editing a book please contact Caroline Mitchell at caroline.mitchell@ioppublishing.org.

Electrochemical Energy Storage Devices and Supercapacitors

An overview

Sandeep A Arote
S. N. Arts, D.J. M. Commerce and B. N. S. Science College, Sangamner, Ahmednagar, India

IOP Publishing, Bristol, UK

ISBN 978-0-7503-3103-6 (ebook)
ISBN 978-0-7503-3101-2 (print)
ISBN 978-0-7503-3104-3 (myPrint)
ISBN 978-0-7503-3102-9 (mobi)

DOI 10.1088/978-0-7503-3103-6

Version: 20211201

IOP ebooks

British Library Cataloguing-in-Publication Data: A catalogue record for this book is available from the British Library.

Published by IOP Publishing, wholly owned by The Institute of Physics, London

IOP Publishing, Temple Circus, Temple Way, Bristol, BS1 6HG, UK

US Office: IOP Publishing, Inc., 190 North Independence Mall West, Suite 601, Philadelphia, PA 19106, USA

Contents

Preface

This book "**Electrochemical Energy Storage Devices and Supercapacitors: An overview**" focuses on the fundamental aspects of electrochemical energy storage devices and supercapacitors. The first chapter provides in-depth knowledge about the current energy-use landscape, the need for renewable energy, energy storage mechanisms, and electrochemical charge-storage processes. It also presents up-to-date facts about performance-governing parameters and common electrochemical testing methods, along with a methodology for result analysis. The second chapter compares supercapacitors and conventional capacitors and provides a discussion of components and their key parameters, together with performance matrixes and the measurable parameters of supercapacitors. Starting with the different components and their specifications for use in energy storage devices, this chapter details the requirements that each component has to meet to achieve its optimal electrochemical performance. It also sheds light on the different types of supercapacitor, along with their working mechanisms and popular materials. The third and fourth chapters extensively explore different classes of electrode material and the challenges and opportunities for future research into supercapacitors as electrochemical energy storage devices. In addition to a performance assessment of the commonly used materials, this book also specifies the features, merits, demerits, and efforts made to enhance the performance of each class of material used for supercapacitors. The electrochemical performances of different classes of material are summarized and tabulated in the respective chapters. Additionally, recent trends in materials development and technological progression are also discussed. As such, this book meets the requirements of engineering and science students/teachers at the undergraduate, post-graduate, and research levels.

Sandeep Arote

Acknowledgements

I offer my deepest sense of gratitude and special affection first and foremost to my parents Akka-Dada for their continuous encouragement during the completion of this work.

I would like to express my deepest love to my wife Darshana for being always by my side, supporting me, and helping during the referencing and proofreading of this work. I would like to extend tremendous thanks to my sisters, Sonali and Sheetal, for their encouragement and unlimited help during the whole work, both emotionally and physically.

It is my pleasure to express my due thanks to Dr Sanjay Malpani, Chairman, and all the staff of the Shikshan Prasarak Sanstha, Professor Insik In, South Korea, Professor Rangnath Pathare, Professor Dr Arun Gaikwad, Principal, Dr Sanjaykumar Dalvi, HoD(Physics) Sangamner College, Sangamner, for their continual and invaluable support and advice.

I would like to extend very special thanks to Mr Yogesh Hase, Mr Abhijiit Landge, Mr Abbas Pathan, Mr Avdhut Landge, Dr Ravindra Tasildar, Dr Anirudhha Mandlik, Dr Maruti Kusmude, Dr Rajendra Waman, Dr Harish Chavan, Professor Pradip Ghule, Professor S P Landge and the staff of the Physics Department for their continued support during the completion of this work.

I am also fortunate to have this opportunity to thank my beloved daughter Shaili and my nephews, Reejul and Sharvil, for their continuous support during the completion of this work.

I am also grateful to the publishers, authors, and other copyright owners who provided consent to utilize their figures, thereby enhancing the quality of this work.

Last, but definitely not least, I would also like to thank the IOP team, especially Ashley Gasque Stec, Senior Commissioning Editor, IOP Publishing, and Robert Trevelyan, for their continuous encouragement and in-depth technical contributions during the completion of this work. Ashley provided much support during the most challenging parts of this work.

Sandeep Annasaheb Arote

Author biography

Dr Sandeep A Arote

Dr Sandeep A Arote is an Associate Professor of Physics at the Sangamner Nagarpalika Arts, D.J. Malpani Commerce and B. N. Sarda Science College (Autonomous), Sangamner, affiliated to Savitribai Phule Pune University (formerly University of Pune), MS, India. He has 15 years' teaching and 10 years' research experience. He has published over 35 research articles in reputed peer-reviewed international journals and also presented his research work at several national and international conferences. His main research areas include: nanostructured materials for energy generation, storage devices (nanostructured solar cells, supercapacitors, and batteries), and sensors. He has co-authored several textbooks for undergraduate physics courses, such as Handbook of Physics, Practical Course in Physics, Solid State Physics, Nuclear Physics, and a booklet on graph plotting techniques. Furthermore, he is a reviewer for several reputed peer-reviewed international journals.

Contributor

Dr Sandeep Annasaheb Arote of the Department of Physics, Sangamner Nagarpalika Arts, D.J. Malpani Commerce and B.N. Sarda Science College (Autonomous), Sangamner –422605, Ahmednagar, M.S. India

Abbreviations

1D	One dimensional
2D	Two dimensional
3D	Three dimensional
AC	Activated carbon
AC-C	Chemically activated carbon
ACFs	Activated carbon fibers
ACNFs	Activated carbon nanofibers
AC-P	Physically activated carbon
BET	Brunauer–Emmett–Teller
CCD	Cyclic charge–discharge
CE	Coulombic efficiency
CNFs	Carbon nanofibers
CNT	Carbon nanotube
CV	Cyclic voltammetry
DFT	Density functional theory
EDL	Electric double layer
EDLC	Electric double-layer supercapacitor
EE	Energy efficiency
EIS	Electrochemical impedance spectroscopy
ESR	Equivalent series resistance
EV	Electric vehicle
GCD	Galvanostatic charge–discharge
G-CF	Graphitic carbon fiber
GDY	Graphdiyne
GNSs	Graphene nanosheets
GO	Graphene oxide
rGO	Reduced graphene oxide
GQDs	Graphene quantum dots
H-CF	Hybrid carbon fiber
HGA	Holey graphene aerogel
HOMO	Highest occupied molecular orbital
HPC	Hierarchical porous carbon
HSC	Hybrid supercapacitor
IEA	International energy agency
IHP	Inner Helmholtz plane
IL	Ionic liquid
ITO	Indium tin oxide
LCO	Lithium cobalt oxide
LHS	Latent heat storage
LISC	Lithium-ion supercapacitor
LMO	Lithium manganese oxide
LNMO	Lithium nickel manganese oxide
LIBS	Lithium-ion battery storage
LTO	Lithium titanate oxide
LUMO	Lowest unoccupied molecular orbital
MOs	Metal oxides
MSs	Metal sulfides

MO@C	Metal-oxide–carbon nanocomposite
MS@C	Metal-sulfide–carbon Nanocomposite
MWCNTs	Multiwalled carbon nanotubes
NSs	Nanosheets
NHGA	Nitrogen-doped holey graphene aerogel
OHP	Outer Helmholtz plane
PSC	Pseudo-supercapacitor
SC	Supercapacitor
SEM	Scanning electron microscopy
SHE	Standard hydrogen electrode
SHS	Sensible heat storage
SSA	Specific surface area
SWCNTs	Single-walled carbon nanotubes
TPES	Total primary energy supply
TMOs	Ternary metal oxides
TNS	Tin nickel sulfide
VE	Voltage efficiency

Chemical elements/compounds and their names/formulae

Ag	Silver
Al	Aluminium
Al_2O_3	Aluminium oxide
$AlPO_4$	Aluminium phosphate
Ar	Argon
Au	Gold
B	Boron
B_2O_3	Boron trioxide
Bi_2O_3	Bismuth oxide
Bi_2S_3	Bismuth sulfide
CA	Cellulose acetate
C-CNFs	Composite carbon nanofibers
$CdMn_2O_4$	Cadmium manganese oxide
$CoNiP_2O_7$	Cobalt nickel pyrophosphate
CH_3COOK	Potassium acetate
$(C_2H_5)4N(BF_4)$	Tetraethylammonium tetrafluoroborate
$CNF@CoNi_2S_4$	Carbon nanofiber–cobalt nickel sulfide nanocomposite
Co	Cobalt
CO_2	Carbon dioxide
(Co_2S_3, Co_3S_4, Co_4S_3 and Co_9S_8)	Cobalt sulfide
Co_3O_4	Cobalt tetroxide
$Co_3V_2O_8$	Cobalt vanadium oxide
$CoMoO_4$	Cobalt molybdate
$CoMoS_4$	Cobalt molybdenum sulfide
$CoNi_2S_4$	Cobalt nickel sulfide
CoO	Cobalt oxide
CoS@CNT	Cobalt sulfide–carbon nanotube nanocomposite
CoS_2	Cobalt sulfide
$CoS_2@C$	Cobalt sulfide–carbon nanocomposite
CuS	Copper sulfide

Cu_2NiSnS_4	Copper nickel tin sulphide
$CuCo_2S_4$	Copper cobalt sulfide
$CuCo_2S_4$@G	Copper cobalt sulfide–graphene nanocomposite
CuO	Copper oxide
CuO@MnO_2	Copper oxide–manganese dioxide nanocomposite
CuO@NiO	Copper oxide–nickel oxide nanocomposite
CuS@CNT	Copper sulfide–carbon nanotube nanocomposite
$CuSbS_2$	Copper antimony sulfide
(Fe_2O_3, Fe_3 O_4)	Iron (III) oxide or ferric oxide
$FeCl_3$	Ferric chloride
$FeNi_2S_4$	Iron–nickel sulfide (violarite)
FeS_2	Iron disulfide
FeS_2@C	Iron disulfide–carbon nanocomposite
GeS	Germanium disulfide or germanium (IV) sulfide
H_2O_2	Hydrogen peroxide
H_2SO_4	Sulphuric acid
H_3PO_4	Phosphoric acid
HNO_3	Nitric acid
IrO_2	Iridium oxide
K_2CO_3	Potassium carbonate
KCl	Potassium chloride
$KMnO_4$	Potassium permanganate
KOH	Potassium hydroxide
$Li_4Ti_5O_{12}$	Lithium titanate
$LiClO_4$	Lithium perchlorate
$LiCoO_2$	Lithium cobalt oxide
$LiFePO_4$	Lithium iron phosphate
$LiMn_2O_4$	Lithium manganese oxide
$LiNi_{0.5}Mn_{1.5}O_4$	Lithium nickel manganese oxide
$LiNi_{1/3}Co_{1/3}Mn_{1/3}O_2$	Lithium nickel cobalt manganese oxide
$LiNbO_3$	Lithium niobate
Li_2MnSiO_4	Lithium manganese silicate
$LiPF_6$	Lithium hexafluorophosphate
LiPON	Lithium phosphorus oxynitride
Mg	Magnesium
MgO	Magnesium oxide
Mn	Manganese
$MnCo_2O_4$	Magnesium cobaltite
$MnCo_2S_4$	Manganese cobalt sulfide
$MnFe_2O_4$	Manganese iron oxide
$MnMoO_4$	Manganese molybdenum tetraoxide
MnO_2	Manganese oxide
MnS	Manganese sulfide
MnS_2@CNT	Manganese (II) Sulfide–carbon nanotube nanocomposite
Mo	Molybdenum
MoO_3	Molybdenum trioxide
MoS_2	Molybdenum disulfide
MoS_2@C	Molybdenum disulfide–carbon nanocomposite
MoS_2@G	Molybdenum disulfide–graphene nanocomposite

N	Nitrogen
Na_2CO_3	Sodium carbonate
Na_2SO_4	Sodium sulfate
$NaClO_4$	Sodium perchlorate
$NaHCO_3$	Sodium bicarbonate
$NaNH_2$	Sodium amide
NaOH	Sodium hydroxide
Nb_2O_5	Niobium pentoxide
$Nb_{14}W_3O_{44}$	Niobium tungsten oxide
N-C/CNT	N-doped carbon-coated carbon nanotube
N-CNFs	Non-cross-linked carbon nanofibers
NH_4Cl	Ammonium chloride
Ni	Nickel
$Ni(OH)_2$	Nickel hydroxide
(Ni_3S_4, Ni_6S_5, Ni_7S_6, Ni_9S_8)	Nickel sulfide
Ni_3S_2@MoS_2	Nickel sulfide–molybdenum disulfide nanocomposite
$NiCo_2O_4$	Nickel cobaltite
$NiCo_2S_4$	Ternary nickel cobalt sulfide
$NiCo_2S_4$@C	Ternary nickel cobalt sulfide–carbon nanocomposite
$NiMoO_4$	Nickel II orthomolybdate
NiO	Nickel (II) oxide
NiOx@CNT	Nickel oxide–carbon nanotube nanocomposite
NiS@C	Nickel sulfide–carbon nanocomposite
NiS@CNT	Nickel sulfide–carbon nanotube nanocomposite
O_2	Oxygen
P	Phosphorous
PAN	Polyacrylonitrile
PbS@CNT	Lead sulfide–carbon nanotube nanocomposite
PC	Propylene carbonate
Pt	Platinum
Ru	Ruthenium
RuO_2	Ruthenium dioxide
S	Sulphur
Si	Silicon
SiO_2	Silicon dioxide
$SnNi_2S_4$	Tin nickel sulfide
SnO_2	Tin oxide
SnS_2	Stannic sulfide, tin sulfide, tin disulfide
Sr	Strontium
$TEABF_4$	Tetraethylammonium tetrafluoroborate
$TEMABF_4$	Triethylmethylammonium tetrafluoroborate
Ti	Titanium
TiO_2	Titanium oxide
$TiNb_2O_7$	Titanium niobium oxide
V	Vanadium
V_2O_5	Vanadium pentoxide
VO_2	Vanadium dioxide
VS_2	Vanadium disulfide
WO_3	Tungsten trioxide
WO_3@CuO	Tungsten trioxide–copper oxide nanocomposite

WS_2	Tungsten disulfide
Zn_2SnO_4	Zinc stannate
$ZnCl_2$	Zinc chloride
$ZnCo_2O_4$	Zinc cobaltite
$ZnCo_2S_4$	Zinc cobalt sulfide
ZnS	Zinc sulfide
ZnS@G	Zinc sulfide–graphene nanocomposite
Zr	Zirconium

Symbols

m	Mass
T	Temperature
P	Pressure
t	Time
V	Volume
V_p	Pore volume
A	Area
e	Charge
n	Electron density
p	Hole density
z	Number of transferred electrons per mole
v	Velocity
σ	Electrical conductivity
I	Current
I_0	Magnitude of sinusoidal current
J	Current density
J_n	Current density of electrons
J_p	Current density of holes
μ_e	Electron mobility
μ_p	Hole mobility
v_d	Drift velocity
μ	Electrochemical potential
μ_A	Chemical potential of the anode
μ_C	Chemical potential of the cathode
G	Gibbs free energy
ΔG	Change in Gibbs free energy
∂G_i	Change in free energy
∂N_i	Number of particles
$\mathbf{E}$	Strength of electric field in volts per meter ($V\ m^{-1}$) or newtons per coulomb ($N\ C^{-1}$)
E_0	Constant amplitude of electric field
D	Diffusion coefficient for charged particle
D_e	Diffusion coefficient for electrons
D_p	Diffusion coefficient for holes
$\frac{dC}{dx}$	Concentration gradient in the x direction
k_B	Boltzmann constant
F	Faraday constant ($F = 96485\ C\ mol^{-1}$)
R	Ideal gas constant ($R = 8.3144598\ J\ mol^{-1}$)
ϕ	Electric potential
$\frac{d\Phi}{dx}$	Potential gradient
$\mathbf{c}$	Concentration
U	Internal energy
H	Enthalpy
ΔH	Change in enthalpy
S	Entropy
ΔS	Change in entropy
$\mathbf{C}$	Capacity

M	Molecular weight
Q	Faradaic charge
dV	Potential difference
ΔV	Potential window
$\frac{dV}{dt}$	Scan rate in mV s^{-1}
C	Capacitance
V_{oc}	Open-circuit voltage
C_s	Specific capacitance
E_d	Energy density or specific energy
P_d	Power density or specific power
P_{loss}	Power loss
ESR	Equivalent series resistance
η	Coulombic efficiency
ω	Angular frequency ($\omega = 2\pi f$)
φ	Phase angle
λ	Debye length
δ	Nernst diffusion-layer thickness
$tan(\varphi)$	Dielectric loss tangent
R_{ct}	Charge transfer resistance
Z	Total impedance
Z'	Real components of Z
Z''	Imaginary components of Z
τ_{max}	Time constant corresponding to a maximum of Z'' in the Nyquist plot
τ_{min}	Time constant corresponding to a maximum of $tan(\varphi)$
Z_w	Warburg impedance
I_D	Diffusion length

Units

g	Gram
s	Second
m	Meter
m^2	Square meter
nm	Nanometer
Pa	Pascal
TPa	Terapascal
V	Volt
mV	Millivolt
A	Ampere
mA	Milliampere
C	Coulomb
Hz	Hertz
MHz	Megahertz
KWh	Kilowatt hour
MWh	Megawatt hour
Wh kg^{-1}	Watt-hour per kilogram
Wh L^{-1}	Watt-hour per liter
W L^{-1}	Watt per liter
W m^{-1} K^{-1}	Watt per meter kelvin
Ah	Ampere hour
Ah g^{-1}	Ampere hour per gram
mAh g^{-1}	Milliampere hour per gram
A g^{-1}	Ampere per gram
mA g^{-1}	Milliampere per gram
A m^{-2}	Ampere per square meter
mA cm^{-2}	Milliampere per square centimeter
g mol^{-1}	Gram per mole
g cm^{-3}	Gram per cubic centimeter
m^2 g	Square meter gram
m^2 g^{-1}	Square meter per gram
cm^3 g^{-1}	Cubic centimeter per gram
cm^2 V^{-1} s^{-1}	Square centimeter per volt second
W kg^{-1}	Watt per kilogram
KW kg^{-1}	Kilowatt per kilogram
V m^{-1}	Volt per meter
V cm^{-1}	Volt per centimeter
V cm^{-2}	Volt per square centimeter
V s^{-1}	Volt per second
mV s^{-1}	Millivolt per second
S cm^{-1}	Siemens per centimeter
S m^{-1}	Siemens per meter
C g^{-1}	Coulomb per gram
F g^{-1}	Farad per gram

$N\ C^{-1}$	Newton per coulomb
$J\ mol^{-1}$	Joule per mole
$C\ mol^{-1}$	Coulomb per mole
ppm	Parts per million
$J\ K^{-1}$	Joule per kelvin

Chapter 1

Electrochemical energy storage mechanisms and performance assessments: an overview

The current practice of researchers working in the area of electric energy generation is to focus on the development of technologies for the utilization of clean, abundant, and cost-effective renewable resources. At the same time, vast progress has been made in the development of energy storage devices with improved cycle life, energy, and power density. In this context, electrochemical energy storage devices have drawn the attention of researchers and industrialists, due to their long cyclic stability and scope for versatile designs using various nanostructured materials. However, a fundamental understanding of their working principles, electrochemistry, key parameters, and performance assessment techniques is essential. This chapter gives an overview of the current energy landscape, energy storage techniques, fundamental aspects of electrochemistry, reactions at the electrode surface, charge conduction and storage mechanisms, factors governing the electrochemical energy storage capabilities of electrodes, electrochemical performance-governing parameters, and electrochemical testing methods. Finally, it offers a comparison of electrochemical testing methods from the performance assessment point of view.

1.1 Introduction to the current energy landscape

Today, living a healthy and prosperous life on an Earth in tune with nature is a priority of human society. Among our various basic needs, energy is an essential requirement for health and also for wealth. Revolutionary changes have made life more pleasant through progressive developments in science and technology. Before industrialization, energy from natural waste and mechanical energy were used. Thereafter, as demand has increased exponentially and the domain of energy requirements has expanded, the search for, and utilization of various sources of energy to fulfill all of today's demands have been continual. Since energy is a fundamental input for economic growth, the overall global energy demand is rising

doi:10.1088/978-0-7503-3103-6ch1

daily as a result of worldwide industrialization, and will continue to do so, as the world's population is continuously growing. To date, conventional energy sources, such as fossil fuels and coal, have been the major contributors to the world's crucial energy requirements.

Figure 1.1 illustrates the world total primary energy supply (TPES) according to source. Based on the data updated in July 2020 by the International Energy Agency (IEA), in 2018, about 31.5% of the world TPES was sourced from fossil fuels and about 26.9% was sourced from coal. Although these sources fulfill energy needs, after consumption, which produces hazardous pollutants, there are negative impacts on the environment and human health. Additionally, the price and security of fossil fuels are among the serious issues described as priorities in [1].

Thus, in brief, the use of crude oil and coal increases the pollution level in the environment and contributes to global warming to a significant extent. Not only is there an environmental impact but also the use of these fuels can be intense in a particular area, causing socioeconomic pressures at an international level. So, there is an urgent need to switch to the processes and types of renewable resource that can be used for sustainable energy generation and growth.

1.2 Renewable energy and the need for energy storage devices

One of the challenges in the twenty-first century is to develop technology for clean, economic, and sustainable energy to meet the energy needs of billions of people. However, research into non-polluting, abundant, and efficient energy alternatives needs to be greatly accelerated in order to do so [2, 3].

Renewable energy, in particular, solar energy, is one of the most promising options because of its accessibility and abundance. According to IEA statistics, among all energy sectors, energy from renewable sources experienced the highest rate of generation increase in 2018 [1].

However, due to variations in the availability and accessibility of renewable energy sources (such as wind speed amplitude, direction, continuity, daily solar

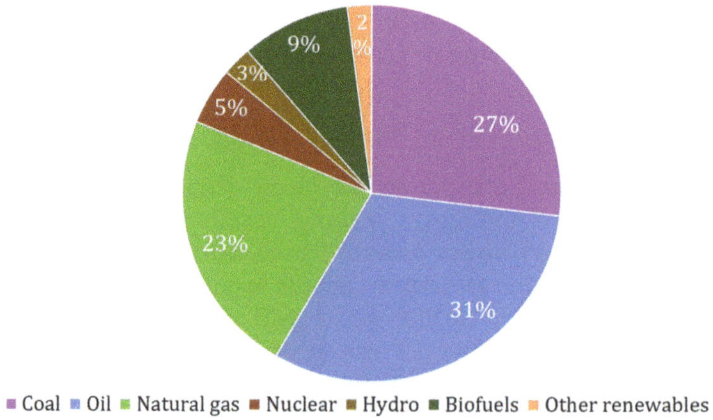

Figure 1.1. World TPES by source in 2018 ([1], accessed on January 06, 2021).

irradiance availability, temperature, etc), the stability of the generated electric energy is affected by fluctuations. After generation, the energy needs to be transported or stored, because, in practice, the ratio of generation to demand may not be 1:1. Transportation though a grid, a conventional method for electric networks, also requires huge investment and infrastructure. In addition, there is a significant disparity between the electric networks of rural and urban areas. To overcome these challenges, the storage of energy by an efficient energy storage device with a long life cycle is one of the best solutions. It is believed that the coupling of renewable energy with efficient energy storage devices will be an effective step toward meeting global energy needs and controlling environmental pollution. Such a hybrid coupling not only provides energy during periods of low sun intensity but also provides grid stability and decentralization for security purposes [4, 5]. This green path not only fulfills clean and sustainable energy demands but also helps the environmental management and socioeconomic development of society. Also, energy storage technologies will help to resolve this issue by the option of on-site distribution and their feasibility [3–6].

1.3 Energy storage techniques

Energy storage devices are designed according to their suitability and capability. Energy can be stored in different forms, such as chemical, mechanical, electric, electrochemical, and thermal. All these forms are discussed in brief in the following section [7–9].

Chemical energy is stored in the structure of a material and depends upon the bonding between atoms or molecules. When a chemical reaction occurs, energy is released, which can be further utilized in the form of electricity or thermal energy. The chemical reaction can be endothermic or exothermic; accordingly, this energy storage mechanism is known as thermochemical energy storage. It is available in different primary forms, such as coal, gas, crude oil, biomass, etc, but due to its hazardous byproducts, it causes environmental pollution.

Mechanical energy is stored by the exchange of kinetic and potential energy. The most common examples are pumped hydroelectric power plants, compressed-air energy storage, and flywheel energy storage.

Electric energy is stored in the form of electric or magnetic energy. It occurs due to the electrostatic behavior of charged particles. Capacitors, induction coils, and electromagnets are examples of electric energy storage devices.

Thermal energy is generally stored as heat energy. It is classified into sensible heat storage (SHS) and latent heat storage (LHS). Liquids that undergo a change in temperature but not in phase (for example, water, mineral oils, etc) are mostly used as media in SHS, while in LHS, the medium also changes its phase with temperature. The most common media for LHS are metals, paraffin, capric acid, etc. Thermal energy is widely used in both industrial and residential applications.

Electrochemical energy is an emerging energy storage class based on the conversion of electric into chemical energy or vice versa. In principle, energy is stored electrochemically via two processes known as the faradaic and non-faradaic processes. The faradaic process is also known as the direct method, in which electric energy is stored by converting it into chemical energy via the oxidation and reduction of an electrochemically active material. The non-faradaic process is also known as the indirect method, in which energy is stored electrostatically by the formation of an electric double layer near the electrode.

Electrochemical cells and systems play a key role in a wide range of industry sectors. At present, these devices are showing their potential for application in diverse sectors, including energy management, conservation, storage, renewable energy, pollution control or monitoring, and greenhouse gas reduction. The different classes of energy storage technique are summarized in figure 1.2.

1.4 Electrochemical energy storage

Electrochemical storage devices, such as Li-ion batteries (LIBs), fuel cells, Li-S batteries, and supercapacitors have great potential to provide increased power and energy density.

Batteries store energy through an electrochemical reaction when charged by an external electric source and can supply electric energy to an external load during discharge; however, in supercapacitors, charges become separated due to the formation of an electric double layer during charging. In fuel cells, the reactants continually supply energy to the electrodes, instead of storing it. Performance-governing parameters, such as the coulombic efficiency, power density, energy density, specific capacitance, capacity retention rate, and cyclic stability need to be optimized to support the use of these devices at the mass level of electrification and for EV technology [10].

Recently, LIBs have demonstrated their prominent candidacy as power supply units for portable electronic devices such as phones, notebook computers, digital cameras, etc. Technologies based on supercapacitors and Li-S batteries are still in the early stages of development for use in several applications that demand high energy and power density, including space and electric vehicles. In spite of the successful commercialization of LIBs, there is still significant scope for improvement in terms of low-cost novel material combinations and safety issues [11]. Before going

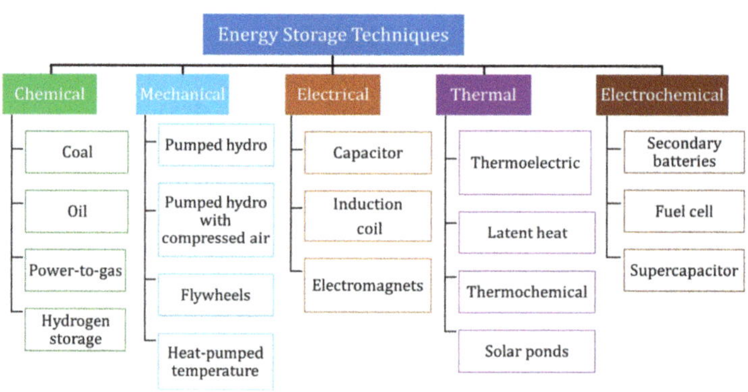

Figure 1.2. Classification of energy storage technologies.

into the details of electrochemical energy storage devices, it is necessary to understand the basics of material electrochemistry, the required material characteristics, and the device performance-governing parameters. The following section discusses all these basic concepts in brief, with some examples.

1.4.1 Electrochemistry: fundamental aspects

The study of chemical reactions that cause electricity by the moment of electrons is known as electrochemistry. It simply describes the study of reactions resulting from chemical and electric processes. Energy released during a chemical reaction can be used to generate electricity (known as electrodics), and electric energy can be used to stimulate a chemical reaction (referred as ionics) [11, 12].

Electrochemical reactions are the fundamental backbone of electrochemical energy storage devices. Most electrochemical devices consist of two electron-conducting electrodes (known as the anode and the cathode) and an ion-conducting electrolyte. During an electrochemical reaction, a moment or exchange of electrons takes place between the electrodes due to a redox reaction at the interfaces between electrodes and the electrolyte. The two electrodes are then connected through an external load to complete the electric circuit [12, 13].

The open-circuit voltage of an electrochemical cell is determined by the difference between the chemical potentials of its electrodes, while the working voltage is defined by the electrochemical window of the electrolyte. In the case of LIBs, the choice of electrode material is decided by comparing the chemical potentials of the highest occupied molecular orbital (HOMO) and the lowest unoccupied molecular orbital (LUMO) levels of the electrolyte [12]. As shown in figure 1.3, for oxidation to take place at the anode and reduction to take place at the cathode, it is necessary for the chemical potential of the anode (μ_A) to lie below the LUMO and the chemical potential of the cathode (μ_C) to be located above the HOMO.

The chemical, electrochemical, and physical properties of an electrode material are significantly influenced by its energy band structure, crystal structure, and

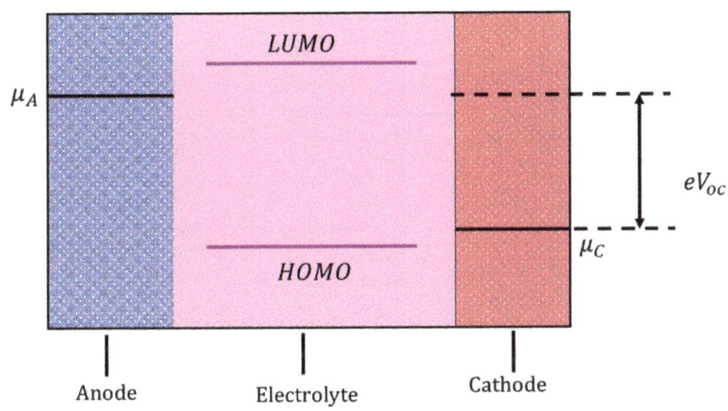

Figure 1.3. Positions of the HOMO and LUMO levels of an electrolyte with respect to the chemical potentials of the electrodes.

microstructure. In addition, the structural parameters can be slightly modified by the insertion of an external dopant. For example, in LIBs, the lithium ions are electrochemically intercalated at suitable cathode lattice sites. A layered structure with sufficient space for intercalation produces high specific capacity and long cyclic stability in electrochemical energy storage devices.

1.4.2 What happens at the electrodes?

The electrochemical reaction that occurs at each electrode is known as a half-cell reaction. Such reactions have two types: the oxidation reaction and the reduction reaction.

Oxidation means the loss of an electron ($M = M^{n+} + ne^-$), which happens at the electrode known as the cathode, while reduction means the gain of an electron ($M^{n+} + ne^- = M$), which happens at the electrode known as the anode. In the case of a supercapacitor, these oxidation and reduction reactions actually happen at the interface between the electrode and the electrolyte [14, 15].

A redox reaction is generally driven by the electrochemical potentials of the anode and cathode materials, as shown in figure 1.4, when two dissimilar metal electrodes (with a large difference in their electrochemical potential) are immersed in an electrolyte and are connected to each other via an external circuit. As a result of electrochemical reactions in the cell, anions (electrons) move from one electrode to the other, while cations pass through the electrolyte. Another important process is electrolysis, in which chemical transformation occurs in a substance because of an electric current passing through it.

Cells are classified into two types, depending upon the electrolysis process. The first is the electrolytic cell, in which chemical changes occur due to the passage of an electric current. The second is the galvanic or voltaic cell, in which chemical reactions result in the generation of electric energy.

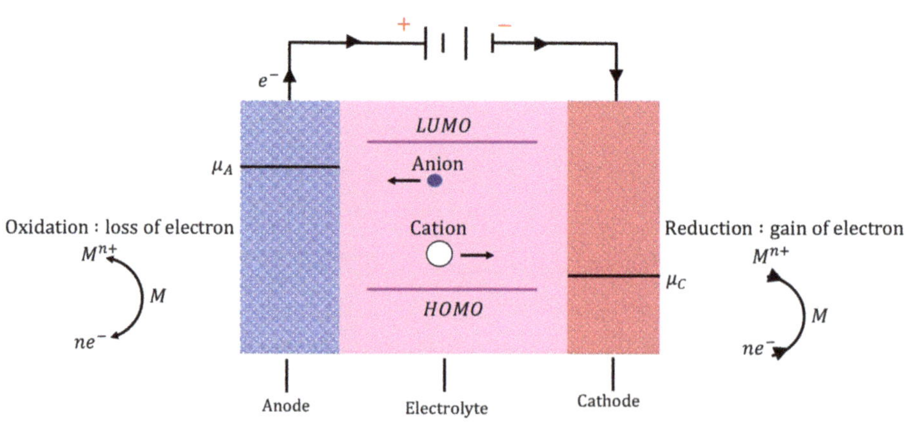

Figure 1.4. Oxidation and reduction processes at the anode and cathode.

1.4.2.1 Faraday's laws of electrolysis

The amount of electric charge transferred and the quantity of substance discharged are correlated by the laws known as Faraday's laws of electrolysis.

First Law: The amount of substance deposited at an electrode is directly proportional to the electric current flowing in the circuit.

Second Law: The masses of the different species deposited at the electrodes by the same electric current are directly proportional to their equivalent weights.

The charge on the electron is 1.6×10^{-19} C. One mole of a substance with one valence electron per atom contains 6.02×10^{23} electrons (Avogadro's number), meaning it has a ~96 500 coulomb charge. This value is known as the Faraday constant. If a current of one ampere is passed through a circuit for one second, then the amount of electricity involved is one coulomb. Thus, the number of coulombs = current in amps x time in seconds. If a 100 mA current passes for ten seconds during the electrolysis of silver nitrate, then the mass of Ag deposited on the cathode will be:

Number of coulombs = 0.10 × 10 = 1 coulomb.

Consider silver and copper, for example:

Let the reaction at the cathode be $Ag^{+} + e^{-} \Rightarrow Ag(s)$.

As stated above, a 96 500 coulomb charge gives one mole of a substance. The atomic weight of silver is 108. Thus, a 96 500 coulomb charge yields 108 g of silver. Therefore, one coulomb yields 108/96 500 = 0.0011g silver on the cathode

If the time is increased to 10 min, then the number of coulombs = 0.10 × 10 × 60 = 60 coulombs.

Sixty coulombs yield (108/96 500) × 60 = 0.067 g of silver on the cathode

Similarly, for the case of Cu, the reaction at the cathode is $Cu^{2+} + 2e^{-} \Rightarrow Cu(s)$.

Thus, two moles of electrons give one mole of copper.

The atomic weight of copper is 63.5 g. Thus, a charge of 2 × 96 500 coulombs gives 63.5 g of copper.

If the time is 10 min and the current is 0.10 A, then the number of coulombs = 0.10 × 10 × 60 = 60 coulombs, which gives (63.5/2 × 96 500) × 60 = 0.0197 g of silver on the cathode.

$$\begin{aligned} &\textit{Mass deposited on cathode during electrolysis (gm)} \\ &= \frac{\textit{atomic weight (gm)}}{\textit{number of valence electrons} \times 96500} \times I\,(A) \times t\,(s) \end{aligned} \tag{1.1}$$

Similarly,

$$\textit{Charge produced} = \frac{\textit{mass deposited (gm)} \times \textit{number of valence electrons} \times 96500}{\textit{atomic weight (gm)}} \tag{1.2}$$

1.4.3 Choice of electrode material and electrolyte

The choice of electrode material and electrolyte in energy storage devices depends on their electrochemical potential, charge capacity, weight, property tunability, and sustainability for a particular application. The electrode material and the electrolyte must possess the required ionic and electronic transport properties along with chemical stability, optimized cost, efficiency, reliability, and lifetime [11]. For example, almost all portable electronic devices (that require energy in the Wh range) currently use LIBs, because Li is light (6.94 g mol^{-1}) and it has the lowest negative electrochemical potential (−3.040 V vs. a standard hydrogen electrode (SHE)) and a low density (0.59 g cm^{-3}). In addition, it also possesses an extremely high theoretical specific capacity (3860 mAh g^{-1}) [16]. To select suitable combinations of materials for high-energy-requirement applications, such as electric vehicles (whose requirements are in the KWh range) and renewable energy (whose requirements are in the MWh range), a deeper knowledge of electrochemistry and electrochemical kinetic processes is necessary.

The following section briefly describes the fundamental aspects of electrochemistry associated with electrochemical storage devices.

1.5 Electrochemical potential

As stated earlier, the electrochemical potential of a chemical species is very important in stimulating an electrochemical reaction. The electrochemical potential is defined as the change in the free energy of a species when an ion or particle is added to it, and is the combined effect of the chemical potential and the local electrostatic potential. In other words, if a change in energy happens between two systems because of their particle exchange (take or give), then it is known as electrochemical potential. The change in free energy is the total of two potentials, chemical and electric, thus it is measured in J mol^{-1} and generally represented by 'μ'.

Sometimes, the chemical potential is referred to as the partial molar Gibbs free energy. It defines the capability of a substance to cause either a chemical or an electrochemical reaction. It is generally responsible for the equilibrium and mass transfer capability of the substance. In any chemical reaction, the driving force of mass transfer is the difference in chemical potential, and the measure of disequilibrium between the reactants and the products is also referred to as the chemical potential. A substance with high chemical potential has low stability and is highly reactive [12, 13, 15].

1.5.1 Electrochemical potential and free energy [16–18]

Let the change in free energy ∂G_i be due to a change in ∂N_i, the number of particles of a species i at some temperature T and a pressure P, then:

$$\mu = \left(\frac{\partial G_i}{\partial N_i}\right)_{T,P}. \tag{1.3}$$

Thus, the unit of chemical potential is the unit of energy. However, the particles of a system are not only atoms or molecules but also electrons, ions, or any others that can be measured. This is why physicists often refer to the Fermi level of an electron as its chemical potential.

The term 'system' usually refers to a 'cell' in electrochemistry. As stated earlier, it consists of two electrodes immersed in an ion-conducting electrolyte. Redox reactions occur in the cell at the electrode–electrolyte interface.

During the redox reaction, ions are exchanged between the two electrodes by the ion-conducting electrolyte, while electrons are exchanged via the external circuit. Such a change happens due to the difference in internal chemical energy, i.e. the chemical potentials of the reactants.

The primary application of an electrochemical cell is to convert chemical energy into electric energy. In many cases, these electrochemical reactions are reversible, and electric energy can be used to convert a chemical substance with low internal chemical energy into a substance with higher internal chemical energy, as is the case in electrolysis.

In an electrolytic cell, a flow of electrons drives nonspontaneous ($\Delta G \geqslant 0$) redox reactions. In a galvanic cell, the energy conversion is achieved by spontaneous ($\Delta G<0$) redox reactions that produce a flow of electrons.

1.5.2 Electrochemical potential and electronegativity [13, 17, 19]

Another important factor that influences the electrochemical potential is electronegativity, which defines the ability of an atom to attract electrons. It increases with atomic number for a given row of the periodic table. Differences in electronegativity are responsible for the formation of chemical bonds. Large differences form ionic bonds, whereas smaller differences form covalent bonds. The type of bond from which a material is formed defines its phase stability and affects the specific site energy available for the intercalation of external species. It also influences the electrochemical potential of the material. Materials formed using ionic bonds have dense structures, while those with covalent bonds have loose or poorly packed structures.

Thus, in short, the redox potential quantifies the energy required to add or remove particles from a system, while the electronegativity quantifies the tendency of a species to gain or lose an infinitesimal fraction of a particle.

The stability and safety, as well as the performance-governing parameters, such as the energy and power densities of electrochemical energy storage devices, are mostly decided by the electronegativity, electron conductivity, ion conductivity, and the structural and electrochemical stabilities of the electrode materials.

1.6 Charge conduction and storage mechanism [20–24]

Along with the electrochemical potential, charge kinetics is also very important for high-performance electrochemical energy storage devices.

The driving force for the motion of charged particles, i.e. electrons or ions, is either an externally applied potential or a concentration gradient present between the species. Charge mobility defines how easily a charged particle can move under the influence of an externally applied field, while diffusivity defines how easily it can move within the concentration gradient between species.

For example, consider the motion of an electron in a current-carrying conductor. According to quantum mechanics, when an external electric field is applied to the conductor, the intrinsic random motion of the electron and its collisions with atoms cause a zig-zag walk of the electron through the conductor; the average velocity with which the electron moves is known as the drift velocity. The speed at which the electron can move is defined by a term known as electron mobility, which is given by:

$$\mu_e = \frac{v_d}{\boldsymbol{E}} \tag{1.4}$$

where μ_e is the electron mobility, v_d is the drift velocity in meters per second (m s^{-1}), and E is the strength of the electric field in volts per meter (V m^{-1}) or newtons per coulomb (N C^{-1}). The drift velocity of electrons in an electric field is proportional to the electric field strength; its direction is opposite to that of E.

From equation (1.4), this can be expressed as:

$$v_d = \mu_e \boldsymbol{E}. \tag{1.5}$$

In the reverse direction, to determine 'v_d' in the presence of an electric field, we consider the effective mass approximation according to the definition of force:

$$m\frac{dv}{dt} = -e\boldsymbol{E}. \tag{1.6}$$

This equation tells us how the velocity changes with time in the presence of an external electric field. Here, the velocity of an electron is the combined effect of both, i.e. its random walk and the electric field.

Upon integration, we obtain:

$$v_d = -\frac{e\boldsymbol{E}t}{m*} \tag{1.7}$$

where t is the average time between collisions.

Thus, from equations (1.4) and (1.7), we obtain:

$$\mu_e = -\frac{et}{m*}. \tag{1.8}$$

The electric current density, J, developed due to the motion of the electron is

$$J = -env \tag{1.9}$$

where e is charge, n is electron density, and v is velocity.

Thus,

$$J = en\mu_e \boldsymbol{E} = \sigma \boldsymbol{E} \tag{1.10}$$

where $\sigma = en\mu$ is the conductivity component contributed by the electrons. Due to drift, the electron also has a diffusion current arising from its random thermal motion (the effects of temperature and entropy) or a difference in chemical potential. Like electric current density, the diffusion current density of an electron can be expressed as:

$$J = eD_e\Delta n \tag{1.11}$$

where D_e is the diffusion coefficient for electrons and Δn is their change in density. If the concentration gradient is only present in one direction, then the one-dimensional diffusion current density is:

$$J_x = eD_e\frac{dn}{dx}. \tag{1.12}$$

Therefore, from equations (1.10) and (1.11), the total current density is a combination of the current densities of both the drift and diffusion mechanisms and is given by

$$J = en\mu_e\boldsymbol{E} + eD_e\Delta n. \tag{1.13}$$

If the applied electric field and the concentration gradients act in the same direction, then

$$J_x(x) = en\mu_e\boldsymbol{E} + eD_e\frac{dn}{dx}. \tag{1.14}$$

In the case of semiconductors, we usually talk in terms of the current densities of electrons and holes, which are as follows:

for electrons,

$$J_n = en\mu_e\boldsymbol{E} + eD_e\Delta n, \tag{1.15}$$

and for holes,

$$J_p = ep\mu_p\boldsymbol{E} + eD_p\Delta p. \tag{1.16}$$

The diffusion coefficient D_e is related to mobility by the Einstein relation:

$$D_e = k_BT\mu_e \tag{1.17}$$

where k_B is the Boltzmann constant.

In most systems, the diffusion and drift currents are complicated by interactions between particles and their proximity to each other, but in electrochemical energy storage devices, the ionic and electronic transports have equal importance. Ionic conduction is different from electronic conduction. The transport of ions through an electrolyte is mainly influenced by the concentration gradient and the electric potential gradient. The concentration gradient causes the diffusion of ions, while the electric potential gradient causes the migration of ions.

In the absence of an electric potential, the total flux of an ion is governed by diffusion and is given by Fick's law, as follows:

$$J_x(x) = D\frac{dc}{dx} \tag{1.18}$$

where D is the diffusion coefficient and $\frac{dc}{dx}$ is the concentration gradient in the x direction.

However, upon the application of an electric field, the ion moves along the potential gradient with a certain effective drift velocity, and the flux is given by the Nernst–Planck law, as follows:

$$J_x(x) = D\frac{dc}{dx}(\textit{diffusion term}) - D\frac{eF}{RT}c.\ \frac{d\phi}{dx}(\textit{Migration term}) \tag{1.19}$$

where $J_x(x)$ is the flux, x is the distance from the surface, e is the charge, c is the concentration, F is Faraday's constant, R is the ideal gas constant, T is the absolute temperature, ϕ is the electric potential, and $\frac{d\phi}{dx}$ is the potential gradient.

1.6.1 Faradaic and non-faradaic processes [20–25]

Two processes take place at the electrodes, namely, the faradaic and non-faradaic processes, which are schematically shown in figure 1.5 [26].

When the accumulation or adsorption of charges at the electrode–electrolyte interface causes a flow of charges in the external circuit, the process involved is known as a non-faradaic process. Here, no charge transfer due to thermodynamic or kinetic processes takes place at the electrode–electrolyte interface; also, there is no change in the solid electrode. The change in electrolyte composition is responsible for the flow of charges. Non-faradaic charge transfer is electrostatic in nature, and the charges are stored by the formation of an electric double layer (EDL).

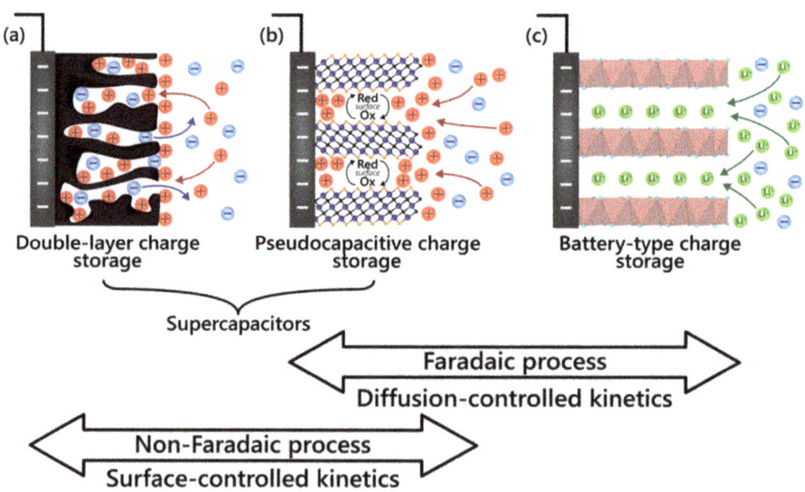

Figure 1.5. Schematic of the faradaic and non-faradaic charge storage mechanisms [26]. John Wiley & Sons. Copyright 2019 Wiley-VCH Verlag GmbH KGaA Weinheim.

However, if oxidation or reduction reactions at the surfaces of the electrodes are responsible for electron transfer, then the process involved is known as a faradaic charge transfer process, because it is governed by Faraday's law. It is also known as a pseudo-capacitive charge transfer.

Electrochemical energy storage devices, such as supercapacitors and rechargeable batteries, work on the principles of faradaic and non-faradaic processes. Supercapacitors use both the EDL and pseudo-capacitive charge storage mechanisms, which means that charges are either stored by the formation of an electric double layer or by a redox reaction at the interface between the electrode and the electrolyte.

However, in the case of batteries, charges are stored by the redox reaction and also by the diffusion of ions (intercalation) in the lattice framework of the electrode materials. Such a storage mechanism is also faradaic, but is non-capacitive in nature. In this case, the electrode undergoes a reversible or quasi-reversible electron transfer reaction. Thus, phase transformation may be present, along with intercalation [26].

In summary, there are two faradaic charge transfer processes: capacitive and non-capacitive. Both happen as a result of the redox reaction at the interface between the electrode surface and the electrolyte; the only difference is that the non-capacitive faradaic charge transfer process is controlled by an ion diffusion mechanism within the crystalline framework of the electrode material.

1.6.2 Factors governing the electrochemical energy storage capability of an electrode

As stated earlier, in order to store energy electrochemically, reservoirs for ions and electrons are needed. Materials with a certain composition, structure, and morphology have the capability to hold charged particles, and hence to store energy electrochemically. Novel designed solid materials with tuned physicochemical and electrochemical properties constitute a priority line of research for efficient electrochemical energy storage. 'Novel design' means an appropriate selection of elements and structure, and optimization of the morphology. For example, the nanostructured forms of a material greatly influence the active surface area, electron transfer pathway, interface properties, and ion diffusion length. The same material with a different morphology may have different electrochemical properties [17, 18].

In electrode materials, chemical energy is stored in the chemical bonds between elements, and is referred to as chemical potential. During an electrochemical reaction, chemical energy is converted into electric energy by the conversion of a material with high bond energy to one with low bond energy. In other words, this phenomenon can be generally described in terms of electrochemical potential, i.e. when a high-electrochemical-potential material is converted into a low-electrochemical-potential material, the chemical energy released is transformed into electric energy. The electrochemical potential of an electrode material depends upon the electronegativity, valence state, ionic radius, electronic structure, crystal structure, and surface–interface structure and bonding [17–19].

The energy change of a given chemical reaction can be described by the Gibbs free energy (G). The change in Gibbs free energy (ΔG) caused by a reaction is the change in the enthalpy (ΔH) of the system minus the product of the temperature and the change in the entropy (ΔS) of the system:

$$\Delta G = \Delta H - T\Delta S \tag{1.20}$$

or, more clearly,

$$\Delta G = \Delta U + P\Delta V + V\Delta P - T\Delta S - S\Delta T \tag{1.21}$$

where

H is the enthalpy (joules), T is the temperature (kelvin), S is entropy (J K^{-1}), U is the internal energy (joules), P is pressure (pascals), and V is volume (m^3).

In electrochemistry, it is represented by the Nernst equation, in which only electric work takes place at a constant temperature, pressure, and volume:

$$\Delta G = -zFE \tag{1.22}$$

where z is the number of transferred electrons per mole in the electrochemical reaction.

Metals with multiple valence states yield high capacities. The theoretical capacity of electrode materials can be expressed by the following equation:

$$C = \frac{zF}{3.6M} \tag{1.23}$$

where M is the molecular weight.

This means that a smaller molecular weight with more valence electrons in an atom can lead to high capacity. Lighter elements, especially those from the first four periods of the periodic table, may yield higher specific capacities because lighter elements have smaller ionic radii and high electronegativity; hence, their valence electrons are bound to the atomic nuclei with more energy, which leads to high electrochemical potential. The factor that influences ionization energy, namely, the electron affinity of an element, has a direct impact on electronegativity and hence changes the electrochemical potential of the material.

For example, lithium has a small molecular weight and multiple valences; as a result, it has a theoretical capacity of about 3850 mAh g^{-1}. Some metals, such as Mn, Fe, Co, Ni, and Cu in their oxide forms exhibit maximum capacity due to a multi-electron transfer reaction [27–30]. For a given metal, the oxide phase, which has a higher electron transfer number, yields high capacity. For example, the theoretical capacity of Cu_2O is 372 mAh g^{-1}, while that of CuO is 674 mAh g^{-1} [30]. Also, the electronegativity difference between metal ions and anionic ligands affects the nature of chemical bond formation, which further affects the crystal structure, its stability, volume density, electrochemical activity sites, and the activation energy required for the insertion of a guest ion. All these factors lead to change in the electrochemical potential of the material, and hence its capacity. In general, closed packed crystalline frameworks in which the tetrahedral and octahedral sites are occupied by cations are favorable for electrodes.

In summary, the preferred electrode materials are elements that have a multiple electron transfer capability, stable layered structures and morphologies with more surface active sites, which are beneficial for electrodes in electrochemical devices.

1.7 Electrochemical performance-governing and measurement parameters

Several factors govern the performance of electrochemical charge or energy storage devices. This section discusses all the technical terminologies/parameters that are used to mention or describe the performance of electrochemical storage devices.

Charge: in electrochemistry, charge is generally referred to as electrochemical charge, which means a charge due to electrochemical processes. It has roughly two categories: *faradaic charge and capacitive charge.*

Faradaic charge is produced by the faradaic process, i.e. the charge transfer process at the interface.

It is given by

$$Q = neN \tag{1.24}$$

where N is the number of reacting species, n is number of electrons, and e is the elementary charge.

In practice, it is detected using the current, I, given by

$$I = \frac{dQ}{dt} = >I = ne.\ \frac{dN}{dt}. \tag{1.25}$$

Thus, the amounts of the reacting species can be calculated by taking the time integral of the current.

Capacitivelnon-faradaic charge is the amount of charge accommodated at the interface. It is independent of the time variation of electroactive species. It is given by

$$Q = C.\ dV \tag{1.26}$$

where C is capacitance and dV is the potential difference at the interface.

Charge density: the amount of electric charge per unit area or volume is known as charge density. In the case of a capacitor, the charge density depends upon the area or volume in between the plates.

The *Lippmann* equation is used to calculate the surface charge density at the interface between an electrode and an electrolyte. According to equation (1.26), the surface charge density can be calculated by taking the ratio of the change in surface tension (the surface energy) to the change in the potential drop across the interface.

The charge density around a specific ion is calculated in terms of the volumetric charge density. This is calculated using the Boltzmann distribution law of classical statistic mechanics or by the electrostatic Poisson equation.

The derivative of the electrode charge density with respect to potential is known as the differential capacity.

Charge transport and charge transfer: these are different concepts. Charge transport is a macroscopic process that takes place over a long distance, compared to charge transfer; examples include the motion of ions in a solution or the conduction of electrons in a metal. On the other hand, charge transfer is a microscopic process that happens at the interface and always results in chemical changes.

Charge transfer occurs at various interfaces, such as the solid–solid, solid–liquid, and liquid–liquid interfaces. At the electrodes, charge transfer occurs homogeneously and also heterogeneously.

If charge transfer process takes place between two similar redox species of the same family, it is known as homogeneous charge transfer.

Open-circuit voltage (V_{oc}): a measure of the difference between the chemical potentials of the negative and positive electrodes of the electrochemical device. The combination of a negative electrode with a high potential and a positive electrode with a low potential yields a high cell voltage.

Capacity: several terms are used to represent the capacity of the electrode material; sometimes, it is represented as a specific capacity, either theoretical or practical.

The total amount of charge stored in an electrochemical device (battery) is known as its capacity; in other words, it is the maximum charge that can be withdrawn from the device. It is expressed in ampere-hours (Ah); 1 Ah is equal to 3600 coulombs. For an electrode, it is calculated by multiplying the number of electrons exchanged by the Faraday constant.

Specific capacity: the amount of charge delivered per unit mass of electrode material. It is expressed in Ah g^{-1} or mAh g^{-1}. This is one of the important parameters used to assess the quality of an electrochemical device.

Theoretical specific capacity: the theoretical specific capacity of any anode or cathode material is calculated by

$$C(Th) = \frac{nF}{3600_{*}M}. \tag{1.27}$$

It is measured in mAh g^{-1}. Here, n is the number of electrons transferred per mole of material, F is the Faraday constant, and M is the molecular weight of the active material used in the electrode.

Theoretical capacity: the theoretical capacity of any electrode material depends upon its theoretical specific capacity and active weight. Here, only the mass of active material is considered, not the masses of the electrolyte, separator, and other components of the device.

Charging capacity: given by the product of the charging time and the current.

Nominal or rated capacity: the maximum constant capacity that can be delivered by the device over several charge–discharge cycles. It depends upon the discharge current, discharge voltage, and the device temperature during operation.

C rate: the current rate at which the electrochemical device charges or discharges. The C rate is based upon the theoretical capacity and the active loading. A device with a capacity of 1 Ah at a rate of 1C will discharge a current of 1 A in 1 h, while a rate of 2 C will discharge it in half an hour.

Equivalent series resistance (ESR): the overall average resistance caused by the combination of the electrode, the electrolyte, the electrode–electrolyte interface, and the electrical contacts. For an ideal electrochemical device, it always equals zero.

Energy density or specific energy (E_d): the amount of energy stored per unit mass or volume, expressed in watt-hours per kilogram (Wh kg^{-1}). It is also known as the gravimetric energy density (energy per unit mass) if represented in Wh kg^{-1} and the volumetric energy density (energy per unit volume) if expressed in watt-hours per liter (Wh L^{-1}).

Power density or specific power (***P_d***): defined as the ability of an electrode active material to generate high currents and voltages per unit of quantity. It is also known as the rate capability of the electrochemical device or the maximum possible discharge rate. It gives the maximum rate of energy discharge or transfer per unit mass or volume. It is also known as the gravimetric power density if represented in watts per kilogram (W kg^{-1}) and the volumetric power density if expressed in watts per liter (W L^{-1}). Electrode materials with high specific areas and mesoporous structures facilitate high power density.

Ragone plot: the balance between energy and power density is very important for the optimum performance of electrochemical devices. For example, supercapacitors have a high specific power density but a small energy density, while secondary batteries offer a good energy density but a much lower power density. The power density and energy density of electrochemical devices are compared using a graph known as Ragone plot. Figure 1.6 shows a Ragone plot for various electrochemical energy storage devices [26]. Using this plot, one can extract an important parameter known as the power-to-energy ratio, which describes the optimum operation of the device.

Efficiency: this defines how efficiently the device performs electrochemical actions. It generally takes three forms: coulombic efficiency (CE), voltage efficiency (VE), and energy efficiency (EE).

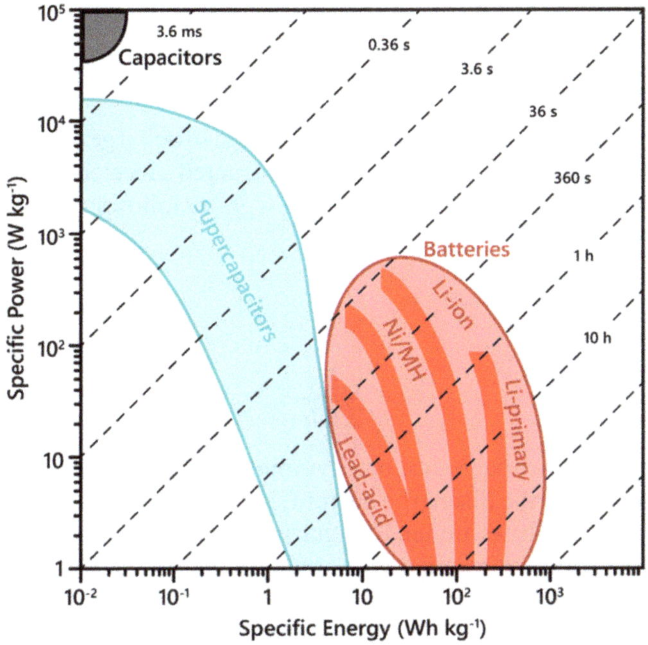

Figure 1.6. Ragone plot presenting the power and energy density correlation for electrochemical charge storage devices [26]. John Wiley & Sons. Copyright 2019 Wiley-VCH Verlag GmbH KGaA Weinheim.

Coulombic efficiency: the efficiency with which stored charge is delivered by the electrochemical device; it is defined as the ratio of the output charge (total charge extracted) to the input charge added to the device over a full cycle. It often improves with cycling. It is also known as the faradaic yield, Faraday efficiency, faradaic efficiency, or current efficiency.

Voltage/voltaic efficiency: the ratio of the average discharging voltage to the average charging voltage of the device. The voltage efficiency depends upon the electrochemical performance of the electrodes. Electrode materials with high active surface areas and high electrical conductivity yield better voltage efficiency.

Energy efficiency: as stated earlier, energy density is nothing more than the total amount of energy that can be extracted from the device, while power density is the rate at which energy can be withdrawn from the device per unit mass or volume. The energy efficiency can be calculated from the ratio of the energy density during discharging to the energy density during charging. In order to improve energy efficiency, the device should work at its optimum energy and power density. Energy efficiency may be preferred as a general metric, but it is unsuitable to be quoted, as it greatly influenced by the charging and discharging rates.

Electric efficiency: this is defined as the product of voltage efficiency and coulombic efficiency. In general, the coulombic efficiency is near 100%, due to very low internal resistance; thus, the voltage efficiency is the only parameter that governs electric efficiency.

The ohmic losses (the resistances of various device components) and electrode polarization loss at the electrodes determine the voltage efficiency of a device.

Cycle efficiency: this is defined as the ratio of energy delivered by a capacitor to the energy supplied to it during a specified cycle.

Charge factor: the reciprocal of coulombic efficiency is known as the charge factor.

Charge–discharge curve: the galvanostatic charge–discharge (GCD) curve is used to analyze the capacitance properties of a constructed electrochemical device. It is also used to calculate the specific capacitance (C_s), as follows:

$$C_s = 4\frac{C}{A} \tag{1.28}$$

with

$$C = \frac{I}{dv/dt} \tag{1.29}$$

where C is the measured capacitance of a symmetrical cell, A is the combined area of the active electrodes, I is the discharge current, and dv/dt is the slope of the discharge curve. The ratio of the areas under the charge–discharge curves multiplied by the current defines the energy efficiency.

Scan rate: this is defined as the step at which the potential of the working electrode changes with respect to that of the reference electrode. For a three-electrode system, it is always measured with respect to the reference electrode. The polarization process is determined by the direction of the scan rate. If the value of the potential w.r.t the reference electrode increases, it is known as the positive direction; otherwise, it is the negative direction. The direction and scan-rate gradient affect the shape of the polarization curve.

Current density: in general, the current density is a measure of the electric current flowing through a conductor per unit of cross-sectional area. It is denoted by J and expressed in terms of A m^{-2}. In electrochemical devices, it is defined as the current capacity per unit area of the electrode. Therefore, it is the current applied or withdrawn from the device. In this case, it is expressed in mA g^{-1}. The product of the charging time (h) and the current density (mA g^{-1}) yields the practical specific capacity (mAh g^{-1}) of the device. The voltage efficiency is strongly affected by the current density, and has inverse an relationship with it. The precise balance between these is a crucial factor for the commercialization of electrochemical devices.

Shelf life: the maximum time for which the electrochemical energy device remains stable is known as its shelf life. If it remains stable, it means that the device works within its specified and acceptable range of performance. It is generally represented in the following form: *'device able to withstand shelf life at ——— °C for ———hours'*.

Cyclic stability: cycle life is a very important parameter for the practical use of electrochemical devices. It defines the capability of the device to complete stable and repetitive charge–discharge cycles by providing a capacity that is more than the rated capacity (lower limit, generally 80%). For all rechargeable devices, it generally depends on factors such as the applied current, voltage limits, and operating temperature.

Supercapacitors and Li-ion batteries: supercapacitors are commonly accepted to be electrochemical energy storage devices that provide high power density, while Li-ion batteries provide high energy density. Supercapacitors have a cycle life two orders of magnitude larger than those of LIBs. The balance between power and energy density is very important for future applications, such as electric vehicles. Supercapacitors work on the principles of non-faradaic and capacitive faradaic charge storage mechanisms, while the operation of batteries depends upon oxidation and reduction reactions in the electrodes. In batteries, operation is based on a non-capacitive faradaic charge storage mechanism, in which the diffusion of ions takes place in an electrode material.

The main challenge is to develop supercapacitors with high energy density while maintaining their inherent characteristics of high power and long cycling life and to develop batteries with high power and long cyclic stability by preserving their high energy density characteristics. Details of the working mechanisms, electrode materials, and recent developments in the field of supercapacitors are given in the respective chapters. The following section describes the commonly employed testing techniques used to monitor the performance of electrochemical energy storage devices.

1.8 Overview of electrochemical testing methods

From the above section, it is very clear that the performance of electrochemical devices can be measured in terms of their specific capacity, energy density, power density, series and parallel resistance, and cyclic stability.

To measure these parameters accurately, several electrochemical methods have been developed, which are trusted by both academia and industry. The testing methods depend on varying the input power parameters, such as the current, voltage, resistance, frequency, and power [31, 32]. The methods used to determine the performance of supercapacitors and batteries, based on the requirements of various applications, include: cyclic voltammetry (CV), the galvanostatic charge–discharge (GCD) or cyclic charge–discharge (CCD) method, electrochemical impedance spectroscopy (EIS), the polarization curve, constant-current charging, potentiostatic tests, etc. In general, a three-electrode test setup known as the half-cell configuration is used for the primary performance measurement of the device. The three electrodes are:

1. the working electrode (the electrode under test);
2. the counter electrode (that completes the circuit);
3. the reference electrode (to provide a constant potential).

The reference electrode provides a constant electrochemical potential. (Commonly used reference electrodes are: the saturated calomel electrode (SCE), the silver/silver chloride (Ag/AgCl) electrode, the mercury/mercury oxide (Hg/HgO) electrode, and the SHE.) In the following subsections, the most commonly used electrochemical performance techniques, CV, GCD, and EIS are discussed in detail.

1.8.1 Cyclic voltammetry (CV) [31, 32]

Cyclic voltammetry is the most widely used electrochemical technique; it measures the change in current in an electrochemical device resulting from a linear sweep of potential within a given range. Figure 1.7 shows the experimental setup used for cyclic voltammetry.

The two control parameters are the voltage window and the scan or sweep rate. During measurement, the voltage is repetitively swept between two extremes. The rate at which the potential changes is known as the scan rate or sweep rate, and is usually represented in mV s^{-1}; the range of potential change is known as the operating potential or potential window. A pair of sweeps in opposite directions is called a cycle.

The electric potential is varied between the positive and negative electrodes of the device, or between the reference and working electrodes for the three-electrode configuration, and the corresponding change in the current passing through the electrochemical is measured. The current is plotted versus the potential, which is known as a CV plot or cyclic voltammogram.

Let $\frac{dV}{dt}$ be the scan rate of the voltage ramp; the current I is then given by

$$I = C.\frac{dV}{dt}. \tag{1.30}$$

The scan rate of the voltage ramp is usually expressed in mV s^{-1}, and it is generally in the range from 0.1 mV s^{-1} to 1 V s^{-1}. The shape of the CV curve is not only

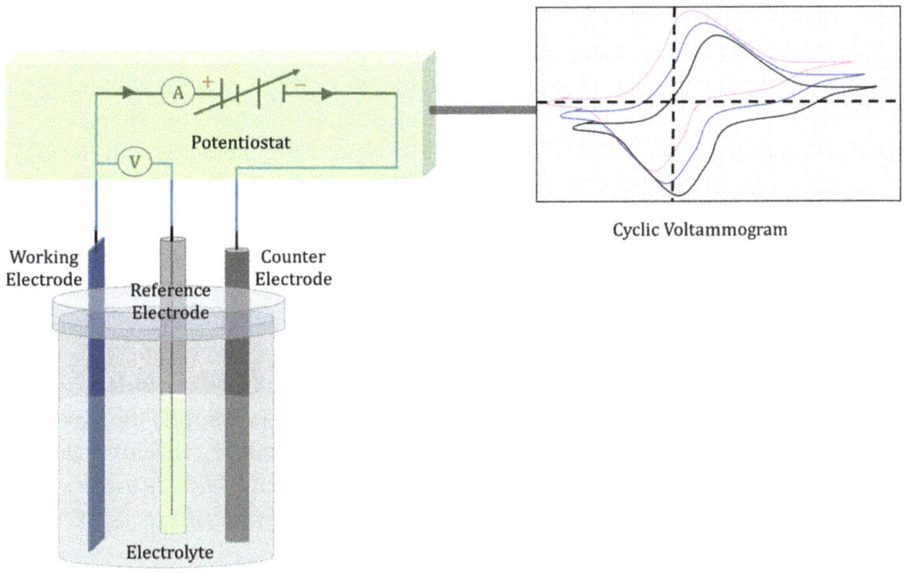

Figure 1.7. Experimental setup used for CV.

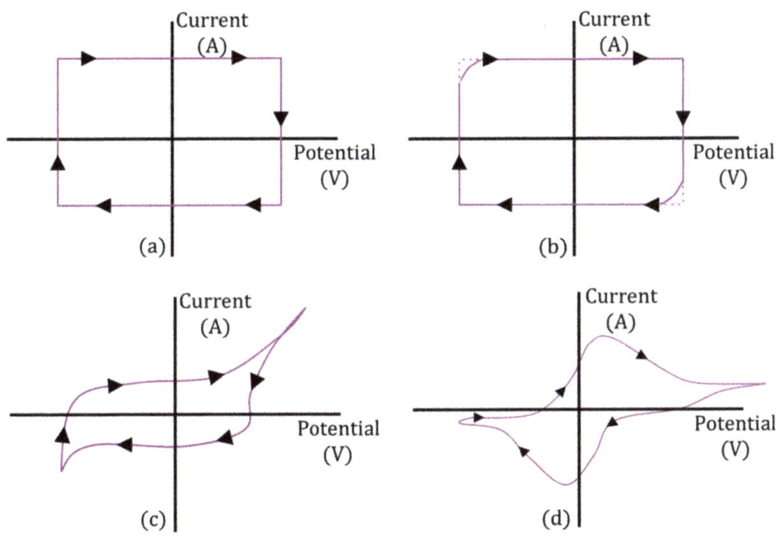

Figure 1.8. Cyclic voltammograms of: (a) an ideal capacitor, (b) a real capacitor, (c) an EDL supercapacitor, and (d) a pseudo-supercapacitor or battery.

dependent on the electrochemical storage device but also on the internal resistance and the scan rate.

Figure 1.8 shows four different shapes of CV curve for electrochemical charge storage devices. For an ideal capacitor (figure 1.8(a)), the shape of CV curve is a perfect rectangle, due to the negligible or null influence of resistance. For a real capacitor (figure 1.8(b)), the curve is different because series resistance plays a key role, causing deviation of the CV curve from its ideal rectangular shape. The magnitude of the current is constant for a linear sweep, positive for an increase in voltage, and negative for a decrease in voltage. However, at the limit potential, where the direction of the voltage changes, the current cannot respond quickly, causing the corners of the CV plot to be rounded off. This type of round-off becomes more prominent with increasing sweep rates. Depending on the charge storage mechanism, electrochemical devices show different CV curve shapes. For example, in case of an electric double-layer capacitance, the electric double layer appears at the interface between the electrode and the electrolyte; in this case, the shape seems to be non-rectangular without peaks (figure 1.8(c)). However, for pseudocapacitance, the CV curve comprises multiple peaks, corresponding to different reversible redox reactions (figure 1.8(d)).

The user has to provide the initial potential, the final potential, and two vertex potential values. CV analysis is based on shape of the curve, the area under the curve, the maximum current values, the peak location, etc. One can also find the operating voltage for a given electrode material, and by integrating the CV curve, one can find the specific capacitance of the device.

1.8.1.1 Calculations based on CV analysis

By calculating the area under the curve of a voltammogram, one can easily estimate the total charge, because the charge is proportional to the surface area under the curve, given by:

$$Q = \int I dt. \tag{1.31}$$

Thus,

$$Charge = \frac{Integrated\ area\ of\ CV}{2 \times Scan\ rate} \tag{1.32}$$

$$Capacitance = \frac{Charge}{Potential\ window} => C = \frac{Q}{\Delta V} \tag{1.33}$$

$$Specific\ capacitance = \frac{Capacitance}{Mass\ of\ active\ material} => C_s = \frac{C}{m} \tag{1.34}$$

and

$$Cs = \frac{Integrated\ area\ under\ the\ curve}{2 \times mass\ of\ electrode \times potential\ window \times scan\ rate} => \\ C_s = \frac{\int I dV}{2 \times m \times \Delta V \times \upsilon} \tag{1.35}$$

where υ is the sweep or scan rate. Using equation (1.34), one can calculate the specific capacitance from CV measurements.

The analysis process will be discussed in detail in the respective chapter, according to the storage class/device.

1.8.2 Galvanostatic charge–discharge (GCD) [31, 32]

The working principle of GCD is the exact opposite of that of cyclic voltammetry. In GCD, the device is charged and discharged between two specified voltages at a constant applied current. GCD is also known as cyclic charge–discharge (CCD) or the chronopotentiometry technique. It is the most popular and trusted method for calculating the specific capacitance and cyclic stability of an electrochemical device under controlled current conditions.

Figure 1.9 shows the GCD plot for four cases. Ideally, the GCD curve (voltage vs. time) should be linear with a positive slope during charging and a negative slope during discharging (figure 1.9(a)). Deviation from the ideal behavior in a GCD plot can be observed in figure 1.9(b), as in the CV curve, which arises from series resistance and the redox reactions. The influence of series resistance on the GCD curve is prominent during the changeover from charging to discharging, and such a deviation from linear behavior is known as an IR drop [32]. Furthermore, the redox reactions are reflected by plateaus in the GC curve (figures 1.9(c) and (d)).

In the three-electrode configuration, a controlled current is applied to the working electrode under test, and the resulting potential is measured with respect to the reference electrode used in this configuration. The most crucial factor for consistent results is the choice of an appropriate level of the applied constant current. In the results, the charging and discharging current is often called the 'C' rate, and the data for cyclic stability are accompanied by a description of the voltage range. For example, in the case of secondary batteries, a common C rate is about 0.1 C, which

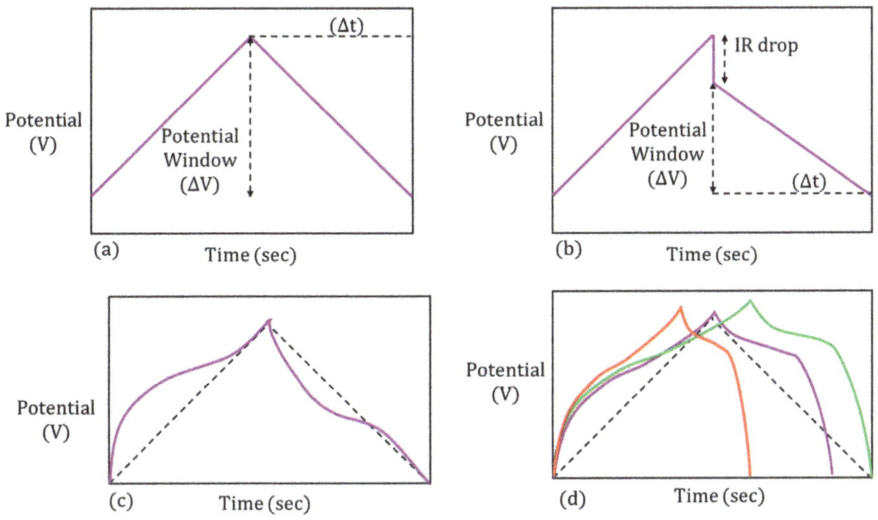

Figure 1.9. GCD plots of: (a) an ideal capacitor, (b) a real capacitor, (c) an EDL supercapacitor, and (d) a pseudo-supercapacitor or battery.

means that the necessary current is fed (drained) to charge (discharge) the device in about 10 h. The cycle life depends on the applied current, voltage window, and temperature. In general, from 5000 to 10,000 repeated cycles are used to report cyclic stability.

1.8.2.1 Calculations based on GCD analysis

In GCD, a controlled current pulse is applied and the resultant potential is measured or plotted as a function of time. The current density is given by

$$Current\ density\left(\frac{A}{g}\right) = \frac{Applied\ current\ (A)}{Amount\ of\ electrode\ material\ (g)}.$$

The specific capacitance is given by

$$C_s = \frac{I}{\Delta V}\frac{\Delta t}{m} \quad (1.36)$$

where I is the constant current applied during discharge, Δt is the discharge time after the IR drop, ΔV is the discharge potential window after the IR drop, and m is the mass of the active material of the working electrode.

Given the GCD data, the energy density (E_d), power density (P_d), ESR, power loss (P_{loss}), coulombic efficiency (η), and specific capacitance can be calculated as follows:

Energy density (E_d):

$$E_d = \frac{C_s \times \Delta V^2}{2} \times \frac{1000}{3600}. \quad (1.37)$$

The multiplication factor is present so that E_d can be presented in Wh kg^{-1}.

Power density (P_d):

$$P_d = \frac{I\Delta V}{2m} \times 1000 => P_d = \frac{E_d}{\Delta t} \quad (1.38)$$

ESR:

$$ESR = \frac{IR_{drop}}{2\,I}. \quad (1.39)$$

Hence. a larger charge/discharge current usually gives rise to a smaller ESR.

Power loss (P_{loss}):

$$P_{Loss} = ESR \times I^2. \quad (1.40)$$

During charge or discharge under extreme conditions, power loss can damage the device.

Columbic efficiency (η):

$$\eta = \frac{I_{Discharge}}{I_{Charge}} \times 100. \quad (1.41)$$

1.8.3 Electrochemical impedance spectroscopy (EIS)

The performance of a realistic electrochemical device is strongly influenced by the ESR. A quantitative analysis of the ESR can be performed using a versatile technique known as EIS. In this technique, a low-amplitude alternating voltage is applied to the device, and the corresponding changes in the amplitude and phase of the current are measured as functions of the frequency [31, 34]. A wide range of frequencies results in a huge amount of data. It is helpful to perform an analysis of the impedance parameters at various interfaces of the device. Figure 1.10 shows the experimental setup used for EIS.

The results are described in the form of the two plots shown in figure 1.11. The first, which represents the variation of the imaginary component of impedance (Z'')

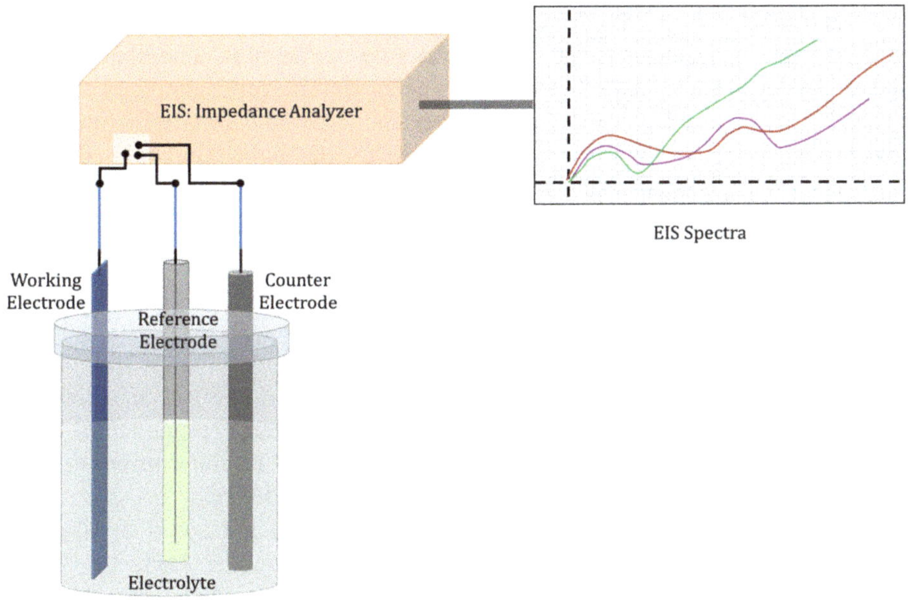

Figure 1.10. Experimental setup used for EIS.

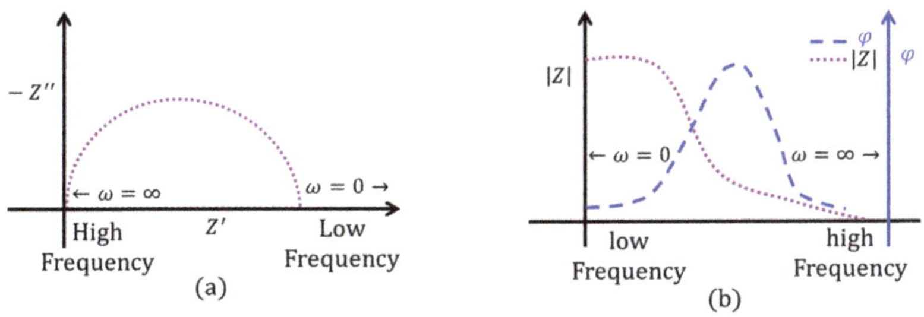

Figure 1.11. Electrochemical impedance spectroscopy represented by: (a) a Nyquist plot and (b) a Bode plot, where $\omega = 2\pi f$ is the angular frequency, φ is the phase angle, and Z' and Z'' are the real and imaginary components of the total impedance Z, respectively.

against the real component (Z'), is known as a 'Nyquist plot' (figure 1.11(a)), while the second, which plots the magnitudes of the impedance and the phase angle against the applied frequency, is known as a 'Bode plot' (figure 1.11(b)). In practice, due to the wide range of frequencies, representation in the form of the logarithm of the applied frequency is preferred for Bode plots [31, 34, 35].

A detailed analysis of these plots helps us to estimate the reaction kinetics, charge transfer resistance, series resistance, electrolyte resistance, capacitance, energy, and power. Such an analysis also helps us to construct an equivalent circuit with different impedance parameters, which further helps us to estimate the contribution of each individual component to the impedance properties.

For example, consider a typical EIS Nyquist plot for a secondary battery, as shown in figure 1.12. Depending on the frequency, the plot is divided into three sections: migration (electrode resistance), charge transfer, and diffusion. The resistive characteristics of the battery can be estimated from the migration part, which is at high frequencies on the left-hand side of the plot. The charge transfer kinetics can be evaluated using the charge transfer section, represented by the middle semicircle in the plot. The diffusion characteristics can be estimated from the low-frequency part, which approaches a vertical line on the right-hand side of the plot, the slope of which gives information about the pore size distribution profile of the working electrode [34, 35].

1.8.3.1 Calculations based on EIS analysis

An electronic circuit with a DC potential satisfies Ohm's law, and the simple resistance of the circuit is given by $R = \frac{E}{I}$. When the applied potential is of the sinusoidal type $E(\omega)$ (an AC voltage), the resulting current ($I(\omega)$) is also sinusoidal, but with a difference in phase (φ). The relationship between $E(\omega)$ and $I(\omega)$ is called the impedance $Z(\omega)$, which is given by $V(\omega)/I(\omega)$. Various circuit components, such as resistors, capacitors, and inductors contribute to impedance.

The impedance of an electrochemical device is usually measured by applying a small sinusoidal potential (~5mV) at multiple frequencies (0.1 MHz to 100 kHz)

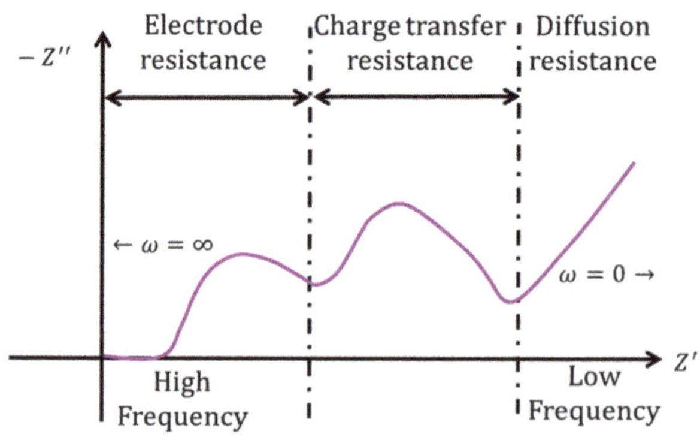

Figure 1.12. Typical EIS spectrum: Nyquist plot for a secondary battery.

between electrodes. The resulting current response is measured to produce an impedance spectrum.

Let the sinusoidal potential be given by

$$\boldsymbol{E} = E_0 \sin(j\omega t) \tag{1.42}$$

where E is the potential at any time t with a constant amplitude of E_0.

The resulting current, with a phase changed by the phase angle φ and with an amplitude I_0 is given by

$$I = I_0 \sin(j\omega t - \varphi). \tag{1.43}$$

According to Ohm's law,

$$Z = \frac{\boldsymbol{E}}{I}. \tag{1.44}$$

Substituting equations (1.42) and (1.43) into equation (1.44) yields:

$$Z(j\omega) = \frac{E_0 \sin(j\omega t)}{I_0 \sin(j\omega t - \varphi)} = |\, Z \,| \exp(j\varphi). \tag{1.45}$$

Thus, φ is the phase angle between the imaginary and real impedances, and

$$Z(j\omega) = |\, Z \,| \exp(j\varphi) = |\, Z \,|(\cos\varphi + j\sin\varphi). \tag{1.46}$$

In practice, for an ohmic resistor,

$$Z = R. \tag{1.47}$$

For a capacitor,

$$Z = \frac{1}{j\omega C}. \tag{1.48}$$

where ω is the angular frequency, which is dependent on the applied frequency:

$$\omega = 2\pi f. \tag{1.49}$$

The total impedance Z is given by

$$\begin{gathered} Z = R + \frac{1}{j\omega C} \quad \text{with} \quad j = \sqrt{-1} \\ => Z = Z' - j \times Z'' \end{gathered} \tag{1.50}$$

where Z' and Z'' are the real and imaginary components of Z, respectively. Thus,

$$Z = \sqrt{(Z')^2 + (Z'')^2} \tag{1.51}$$

with

$$Z' = R \quad \text{and} \quad Z'' = -\frac{1}{\omega .\, C}. \tag{1.52}$$

From equation (1.52), the capacitance can be calculated as follows:

$$C = -\frac{1}{\omega Z''} \Rightarrow C = -\frac{1}{2\pi f Z''}. \tag{1.53}$$

The loss tangent (tan (φ)) of an electrochemical system is given by

$$\tan(\varphi) = \frac{Z'}{Z''} = \frac{\omega\tau_{max}\delta}{1 + \omega^2\tau_{\max}^2\delta} \tag{1.54}$$

where ω is the angular frequency, $\tau_{\max}$ is a time constant that corresponds to the maximum Z'' in the Nyquist plot, and δ (the Nernst diffusion-layer thickness) is the ratio of half of the sample thickness, d, to the Debye length, λ.

The maximum of the loss tangent in the tan(φ)–logf plot appears at

$$\tan(\varphi)_{\max} = \frac{\sqrt{\delta}}{2} at \quad \tau_{\min} = \tau_{\max}\sqrt{\delta} \tag{1.55}$$

where $\tau_{\min}$ is a time constant at which the value tan (φ) attains its maximum in the tan (φ) – logf plot.

Thus, the Debye length is

$$\lambda = \frac{d}{\delta} = \frac{d}{2(\tan(\varphi)_{\max})^2} \tag{1.56}$$

and the diffusion length, I_D, is defined as

$$I_D = \sqrt{(\tau_s \times D)} \tag{1.57}$$

where D is the diffusion coefficient and τ_s is the time constant; τ_s can be calculated from the maxima of the Bode plot.

Thus, the EIS data are represented by two plots, i.e. the Nyquist and Bode plots, the analysis of which gives information about various electrochemical parameters.

1.8.3.2 Modeling of EIS data with basic circuit elements

Electrochemical charge storage devices comprise various interfaces, which are represented by different combinations of circuit elements, known as equivalent circuits. EIS data are further analyzed to represent the system under study using an equivalent circuit. Figure 1.13 shows the EIS plots for various circuit elements and their combinations. As a single component, a resistor follows Ohm's law and is independent of frequency, hence, no phase shift is observed. The Nyquist plot for a resistor shows a single point on the horizontal axis, because $Z' = R$ and $Z'' = 0$, while its Bode plot shows two parallel lines. In the case of an ideal capacitor (figure 1.13(a)), the stored charge is proportional to the applied potential, and the stored charge decays exponentially with time t during discharging. For an AC potential at a high frequency, the current is greater; therefore, the impedance is low and vice versa. The Nyquist plot for a capacitor (figure 1.13(b)) shows a straight line along the vertical axis, while its Bode plot shows a linear plot with a negative slope,

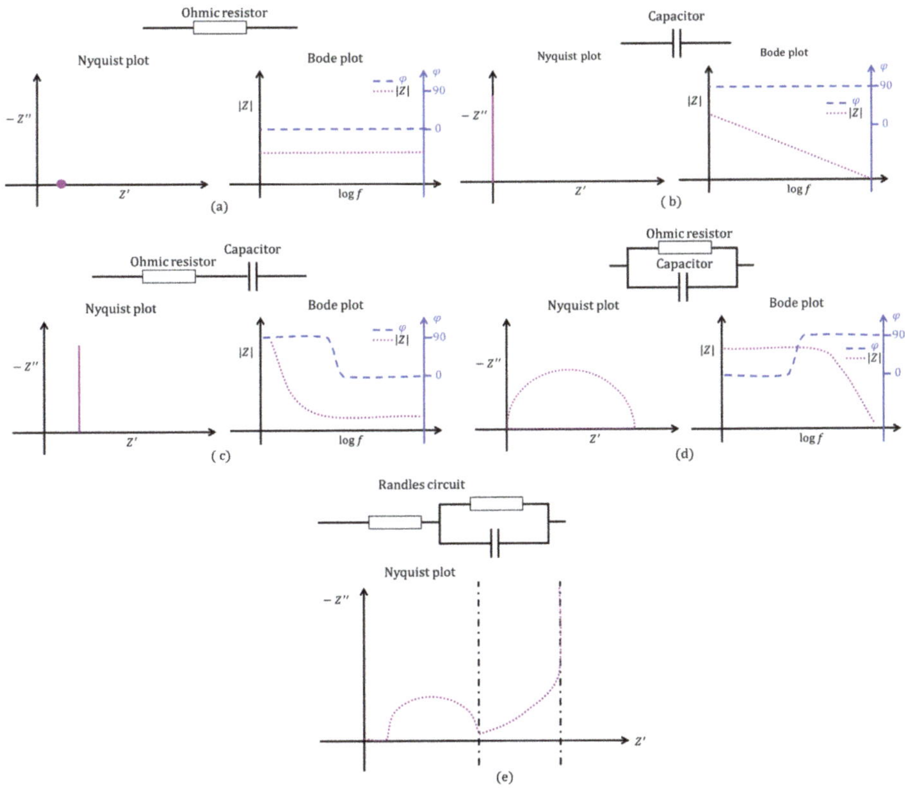

Figure 1.13. Schematic illustration of a Nyquist plot, a Bode plot and the equivalent circuit for: (a) an ideal resistor, (b) a capacitor, (c) their series combination, (d) their parallel combination, and (e) a Randles circuit for series–parallel combinations.

because, according to equation (1.47), at very high frequencies $Z = 0$, and at very low frequencies $Z = \infty$. For an ideal capacitor, $\varphi = 90°$, which is represented by a line parallel to the horizontal axis. Various circuits can be designed with combinations of resistors and capacitors. Figures 1.13(c) and 1.13(d) show EIS plots for series and parallel combinations, respectively. For a series combination, the impedance of a capacitor is greater in the lower frequency range and decreases with an increase in frequency; hence, the overall impact of a capacitor is greater in the low-frequency range. As discussed above, the impedance of a capacitor is smaller at high frequencies and increases with decreasing frequency. So, in the case of a series combination of a resistor and a capacitor, the overall impact of the capacitor is greater in the low-frequency range (figure 1.13(c)). On the other hand, for a parallel combination, the current flows through the capacitor in the high-frequency range and the fraction of the current flowing through the resistor increases with decreasing frequency. Therefore, in the low-frequency range, the imaginary component of the impedance Z'' is low, and with increasing frequency, Z'' also increases, leading to a semicircle in the Nyquist plot (figure 1.13(d)). The radius of the semicircle represents the electrode resistance.

In an electrochemical system, the resistance is composed of electrode resistance, charge transfer resistance, and diffusion resistance. If a charge is moving in the same phase, i.e. within the electrolyte, then such a charge transfer can be represented by a simple ohmic resistor. In case of the interface, the charge either goes from electrode to electrolyte or from electrolyte to electrode, and is represented by the charge transfer resistance (R_{ct}). The Nyquist plot and a simple circuit known as a Randles circuit for these two cases are shown in figure 1.13(e). A more depressed semicircle in a Nyquist plot is sign of lower R_{ct}, while a less depressed semicircle depicts a higher R_{ct} value. The straight line behavior of the curve in the low-frequency region is a representation of the impedance known as Warburg impedance (Z_w); it is a noteworthy indication of ion diffusion at the electrolyte–electrode interface.

Thus, in conclusion, the three electrochemical methods CV, GCD, and EIS are the most reliable and efficient methods for the electrochemical analysis of charge storage devices. The quantitative and qualitative performances of electrochemical devices can be measured by analyzing the results obtained from these methods. A detailed comparison of these methods is presented in table 1.1

1.9 Summary

The rate at which the global energy demand is increasing due to industrialization and globalization has caused energy generation and storage to become a priority for research and also for technology upgrades. The greatest emphasis needs to be given to hybrid technologies that support renewable energy generation, effective storage, and distribution. To date, most of the world's energy needs have been fulfilled using fossil fuels, but these are expensive in terms of socioeconomic impacts and adverse impacts on the environment. Future energy systems will need to produce low levels of carbon emissions, hence, the importance of renewable energy resources has been highlighted in the present era. As a result, novel energy storage approaches are essential for the optimum use of energy. Several energy storage technologies have been proposed, which store electricity using mechanical, thermal, chemical, and electrochemical approaches. Much study has been carried out in this field, but there is room for further discussion about energy storage.

As an introduction, the need for renewable energy, different classes of energy storage technologies, and the importance of electrochemical energy storage have been discussed in this chapter. Electrochemical devices have three major components: the anode, the cathode, and the electrolyte. The best combination of these three is based on the position of the electrochemical potential and the HOMO–LUMO levels of the electrode and the electrolyte, respectively. Several kinematic factors must be optimized while choosing materials and fabricating an electrochemical energy storage device. For example, the energy density, power density, specific capacitance, cyclic stability, cycle life, coulombic efficiency, etc, are the performance-governing parameters that need to be calculated. Among the various electrochemical testing techniques, CV, GCD, and EIS are the most frequently preferred and trusted. Analyzing a CV plot helps us to understand the redox reaction processes and the key role of series resistance. The energy density, power density,

Table 1.1. Summary of the electrochemical (CV, GCD, and EIS) testing methods.

Characteristic	CV	GCD	EIS
Helps to provide information about	• Electrochemical properties of the working electrode • Redox reaction of chemical species on the working electrode • Mechanisms of electron transfer kinetics • Thermodynamic redox processes	• Durability of the electrode material and the stability of the device • Response in terms of potential with respect to time during the charging and discharging of the device • Performance-governing parameters	• Electron transfer kinetics and diffusional characteristics • Analysis of impedance profile at various interfaces in the device • Frequency response of the device • Equivalent circuit of the device
Working mechanism	• The potential is linearly scanned backward and forward between two extreme values at a constant rate (the sweep/scan rate) and the resulting current is measured. • **Scan rate or sweep rate**: the rate at which the potential is swept • **Potential window**: range of the potential change • **Cycle**: A pair of sweeps in opposite directions	• A constant current density is applied and the resulting potential is measured with respect to time. • The device is charged and discharged between two specified voltages at a constant applied current. • **C rate**: charging and discharging current.	• A small sinusoidal potential is applied at multiple frequencies between electrodes and the resulting current response is measured to produce an impedance spectrum. • The impedance and phase shift are measured for a wide range of frequencies (and hence plotted against the logarithmic value of the frequency).
Control parameter	Voltage window and scan or sweep rate	Current density	Frequency
Range	Scan rate :0.1 mV s^{-1} to 1 V s^{-1}	C rate[a]	Sinusoidal potential (~1 to 10 mV). Frequency (~0.1 MHz to 1 MHz).
Output	• The measured current is plotted against a potential that is linearly swept back and forth. • The resulting plot is known as a CV plot or a cyclic voltammogram; it is used to find the optimum potential	• Measured voltage vs. time (positive slope during charging and negative during discharging) • The resulting plot is known as a charge–discharge curve or GCD curve; it is used to study the cyclic	• The results are represented in terms of two plots: • **Nyquist plot**: a plot of imaginary vs real impedance at different frequencies;

(*Continued*)

Table 1.1. (*Continued*)

Characteristic	CV	GCD	EIS
	window for the charge–discharge profile of the device. • The difference between the two peak (cathodic and anodic) potentials of the CV curve gives information about the diffusion rates of the electroactive species	stability of the device within the same potential window (chosen from CV) at different current densities.	• **Bode plot**: absolute impedance and phase vs frequency. • Charge transfer resistance (R_{ct}) at the electrode/electrolyte interface results in a semicircular Nyquist plot in the high-frequency region, and the diffusion of ions from the electrolyte into the electrode, called Warburg impedance (Z_w), results in straight line behavior of the Nyquist plot in the low-frequency region.1pc • Data interpretation is carried out using equivalent electrical circuits.
Calculated device performance parameters	• Specific capacitance[b] • Potential window • Diffusion rate	• Specific capacitance[b] • Energy density • Power density • Coulombic efficiency • Cyclic stability	• Resistance of the electrolyte • Charge transfer resistance • Warburg impedance • Equivalent series resistance (ESR) • Equivalent circuit
Effect of control parameters	• The scan rate has a significant effect on the specific capacitance value. • The specific capacitance decreases with increasing scan rate. • The potential window is limited by the operating stability of the electrolyte solution. • A slow scan rate is preferable.	• The charging and discharging rate affects many parameters. • Fast charging rates result in lower specific capacitance and lower energy density. • A higher current density leads to fast device discharge. • A slow charging rate is preferable.	• The applied frequency range affects the specific capacitance value. • With an increase in frequency, the specific capacitance decreases. • The Warburg element, i.e. the impedance due to the diffusion process, is also dependent on frequency.

[a] The C rate is calculated from the capacity of the device.

[b] GCD cycling usually reports higher specific capacitances than CV; however, if the average current in CV is comparable with the constant current of the linear GCD, then only the specific values calculated from them are comparable. In other words, a sufficiently low scan rate in CV and a sufficiently low current density in GCD can provide comparable capacitance values.

specific capacitance, cyclic stability, cycle life, and coulombic efficiency can be calculated using a detailed analysis of the GCD curves. The ESR contribution to impedance can be judged with help of Nyquist and Bode plots, which are plotted using EIS data. Using the contents of this chapter, the reader can understand the fundamental aspects of energy storage mechanisms and their performance evaluation techniques.

References and further reading

[1] IEA *Total primary energy supply by fuel, 1971 and 2018* IEA, Paris (https://iea.org/data-and-statistics/charts/total-primary-energy-supply-by-fuel-1971-and-2018)

[2] Denholm P, Ela E, Kirby B and Milligan M 2010 Role of energy storage with renewable electricity generation (No. NREL/TP-6A2–47187) *National Renewable Energy Lab.(NREL)* (Golden, CO)

[3] Arent D, Engel-Cox J, Logan J, Zhou E, Ericson S, Anderson K, Cole W, McLaren J, Garfield D and Ruth M 2020 Next generation renewables and, separately, energy system transformation, and related topics: cooperative research and development final report, CRADA Number CRD-14–557 (No. NREL/TP-6A50–76744) *National Renewable Energy Lab.(NREL)* (Golden, CO)

[4] Elshurafa A M 2020 The value of storage in electricity generation: a qualitative and quantitative review *J. Energy Storage* **32** 101872

[5] Hicks W 2020 Declining renewable costs drive focus on energy strong (https://nrel.gov/news/features/2020/declining-renewable-costs-drive-focus-on-energy-storage.html)

[6] Guo Z, Wei W, Chen L, Dong Z Y and Mei S 2020 Impact of energy storage on renewable energy utilization: a geometric description *IEEE Trans. Sustainable Energy* **12** 874–85

[7] Fuchs G, Lunz B, Leuthold M and Sauer D U 2015 Overview of nonelectrochemical storage technologies *Electrochemical Energy Storage for Renewable Sources and Grid Balancing* (Amsterdam: Elsevier)

[8] Amrouche S O, Rekioua D, Rekioua T and Bacha S 2016 Overview of energy storage in renewable energy systems *Int. J. Hydrogen Energy* **41** 20914–27

[9] Hadjipaschalis I, Poullikkas A and Efthimiou V 2009 Overview of current and future energy storage technologies for electric power applications *Renew. Sustain. Energy Rev.* **13** 1513–522

[10] Deng D 2015 Li-ion batteries: basics, progress, and challenges *Energy Science & Engineering* **3** 385–418

[11] Nitta N, Wu F, Lee J T and Yushin G 2015 Li-ion battery materials. present and future *Mater. Today* 18

[12] Liu J *et al* 2013 Materials science and materials chemistry for large scale electrochemical energy storage: from transportation to electrical grid *Adv. Funct. Mater.* **23** 929–46

[13] Chen K and Xue D 2016 Materials chemistry toward electrochemical energy storage *J. Mater. Chem.* **4** 7522–537

[14] Özkaya A R 2002 Conceptual difficulties experienced by prospective teachers in electrochemistry: Half-cell potential, cell potential, and chemical and electrochemical equilibrium in galvanic cells *J. Chem. Educ.* **79** 735

[15] Tan S Y, Chia V Y, Hölttä-Otto K and Anariba F 2020 Teaching the nernst equation and faradaic current through the use of a designette: an opportunity to strengthen key electrochemical concepts and clarify misconceptions *J. Chem. Educ.* **97** 2238–43

[16] Xu W, Wang J, Ding F, Chen X, Nasybulin E, Zhang Y and Zhang J G 2014 Lithium metal anodes for rechargeable batteries *Energy Environ. Sci.* **7** 513–37

[17] Gao J, Shi S Q and Li H 2015 Brief overview of electrochemical potential in lithium ion batteries *Chin. Phys.* B **25** 018210

[18] Newman J and Thomas-Alyea K E 2012 *Electrochemical Systems.* (New York: Wiley)

[19] Gyftopoulos E P and Hatsopoulos G N 1968 Quantum-thermodynamic definition of electronegativity *Proc. Natl. Acad. Sci. USA* **60** 786

[20] Conway B E 1991 Transition from 'supercapacitor' to 'battery' behavior in electrochemical energy storage *J. Electrochem. Soc.* **138** 1539

[21] Liu J, Song X, Zhang T, Liu S, Wen H and Chen L 2021 2D conductive metal–organic frameworks: an emerging platform for electrochemical energy storage *Angew. Chem. Int. Ed.* **60** 5612–24

[22] Fleischmann S, Mitchell J B, Wang R, Zhan C, Jiang D E, Presser V and Augustyn V 2020 Pseudocapacitance: from fundamental understanding to high power energy storage materials *Chem. Rev.* **120** 6738–82

[23] Zheng J P and Jow T R 1995 A new charge storage mechanism for electrochemical capacitors *J. Electrochem. Soc.* **142** L6

[24] Zu L, Zhang W, Qu L, Liu L, Li W, Yu A and Zhao D 2020 Mesoporous materials for electrochemical energy storage and conversion *Adv. Energy Mater.* **10** 2002152

[25] Guan L, Yu L and Chen G Z 2016 Capacitive and non-capacitive faradaic charge storage *Electrochim. Acta* **206** 464–78

[26] Mathis T S, Kurra N, Wang X, Pinto D, Simon P and Gogotsi Y 2019 Energy storage data reporting in perspective—guidelines for interpreting the performance of electrochemical energy storage systems *Adv. Energy Mater.* **9** 1902007

[27] Ohzuku T and Ueda A 1994 Why transition metal (di)oxides are the most attractive materials for batteries *Solid State Ionics* **69** 201–11

[28] Desilvestro J and Haas O 1990 Metal oxide cathode materials for electrochemical energy storage: a review *J. Electrochem. Soc.* **137** 5C

[29] Zhou X, Shi J, Liu Y, Su Q, Zhang J and Du G 2014 Microwave-assisted synthesis of hollow CuO–Cu_2O nanosphere/graphene composite as anode for lithium-ion battery *J. Alloys Compd.* **615** 390–94

[30] Sun L, Deng Q, Li Y, Deng L, Wang Y, Ren X and Zhang P 2016 Solvothermal synthesis of ternary Cu_2O-CuO-RGO composites as anode materials for high performance lithium-ion batteries *Electrochim. Acta* **222** 1650–59

[31] Bard A J and Faulkner L R 2001 Fundamentals and applications *Electrochemical Methods* **2** 580–632

[32] Conway B E 2013 *Electrochemical Supercapacitors: Scientific Fundamentals and Technological Applications* (Berlin: Springer)

[33] Cook D 2014 'Getting started with your first experiment: EIS300 electrochemical impedence techniquies- potentiostatic electrochemical impedance spectroscopy,' *Gamry* 1

[34] Dsoke S, Tian X, Täubert C, Schlüter S and Wohlfahrt-Mehrens M 2013 Strategies to reduce the resistance sources on electrochemical double layer capacitor electrodes *J. Power Sources* **238** 422–29

[35] Nguyen T Q and Breitkopf C 2018 Determination of diffusion coefficients using impedance spectroscopy data *J. Electrochem. Soc.* **165** E826

Chapter 2

Perspectives on electrochemical supercapacitors: working principles and classification

Supercapacitors (SCs), electrochemical devices that use novel nanostructured materials, have attracted much attention recently, due to the possibility that they may provide high power density with optimum energy density in everyday consumer products. Depending on their charge storage mechanisms, they have evolved into different classes: electric double-layer, pseudo, and hybrid SCs. Several models have been developed to describe the charge storage mechanisms of SCs. As a result of continuous development, numerous derived performance-assessment parameters have evolved, which have ensured that SCs are prominent energy storage devices. This chapter discusses the fundamental aspects of charge storage mechanisms in SCs, their components, and their classification. It also familiarizes the reader with the key parameters of the different component materials.

2.1 Introduction

Energy demands are increasing exponentially worldwide, and to meet this demand, various conventional energy sources, such as fossil fuels, coal, etc, are used extensively. These conventional sources are now reaching the threshold of depletion and causing serious environmental impacts. Air pollution caused by vehicles based on fossil fuels is an important issue worldwide. As fossil fuels are no longer sustainable and are concentrated in particular geographical areas, their prices increase continuously. So, there is a need to search for an alternative that will support sustainable, clean, and delocalized energy utilization. Today, renewable energy is a promising alternative to conventional energy sources, but the potential use of renewable energy is possible if and only if it is coupled with efficient energy storage technologies [1–3]. Secondary batteries, a well-known type of energy storage device, are generally used for this purpose. The continuous development of more efficient electrical storage devices has resulted in the emergence of the lithium-ion

doi:10.1088/978-0-7503-3103-6ch2

battery. To date, for most portable electronic devices, Li-ion batteries have proved their worth as efficient energy storage devices [3–5].

However, there are certain areas in which high-power energy is required; for example, the current generation of battery technology has failed to meet the needs of hybrid electric vehicles [6]. The major issues with batteries are low power density, limited cycle life, and heat dissipation [7]. Thus, the development of environment-friendly energy storage devices with high power, high energy, and a long cycle life is a major focus of research and technology.

SCs, also known as electrochemical capacitors, have attracted much attention from academia, research, and industry because of their high power density and long cycle life (>100 000 cycles) [6–9]. Compared to SCs, batteries can store more energy per unit mass or volume, but they deliver it slowly, and hence have lower power densities. SCs, on the other hand, store less energy but deliver it quickly, and hence offer high power densities and rapid discharge [9].

The combination of the high energy density of batteries with the high power density and long cycle life of SCs is suitable for high-energy applications, such as hybrid electric vehicle technologies [10–14].

This chapter gives an overview of capacitors and SCs, including the basic principles of their charge storage mechanisms, their basic structures, and their types. It also discusses the key requirements of each component in SCs and recent developments in electrode materials, such as carbon, carbon derivatives, metal oxides, sulfides, and composites.

2.2 Capacitors and their energy storage mechanisms

A capacitor or condenser is a passive electrical component, characterized in terms of the electrical parameter, capacitance, and used to store electric energy electrostatically.

Capacitance is a well-known term, defined as the ability of an electrode material to store charges electrostatically. It is measured in farads (F), but, in practice, it is generally represented in μF or pF.

The voltage of an ideal capacitor is proportional to the charge stored in it:

$$Q \propto V$$

$$Q = CV \tag{2.1}$$

where C is the capacitance in farads, V is the voltage between the two plates or electrodes, and Q is the stored charge, represented in coulombs or ampere-seconds. However, it is difficult to measure the charge stored at the electrode, so it is generally estimated in terms of current. Let I be the current applied to the electrode, defined as $I = dQ/dt$; equation (2.1) can then be written as,

$$I = C\ \frac{dV}{dt}. \tag{2.2}$$

The power (P) drawn from a capacitor is defined as the rate of transfer of energy. It is given by the product of the current and the voltage of the capacitor. Thus,

$$P = IV => \\ P = C\frac{dV}{dt}V(t). \tag{2.3}$$

But,

$$P = \frac{dE}{dt}.$$

Thus, the energy E stored in a capacitor is given by integrating equation (2.3).

$$E = \frac{1}{2}CV^2 = \frac{1}{2}QV \tag{2.4}$$

where V is the maximum operating voltage of the capacitor and E is the energy in joules.

Theoretically, any electrode material has an intrinsic maximum capacitance, so, ideally, it is expected that a capacitor with such an electrode should achieve that level of charge storage ability. An ideal capacitor has no power or energy loss during charge or discharge.

Figure 2.1(a) shows a schematic of a conventional capacitor with two conductive electrodes separated by a dielectric insulating material. The dielectric material in between the electrodes prevents their direct contact. When voltage is applied to a capacitor, it enables charge to accumulate on the electrode surfaces. The capacitance is given by

$$C = \frac{\varepsilon_0 \varepsilon_r A}{d} \tag{2.5}$$

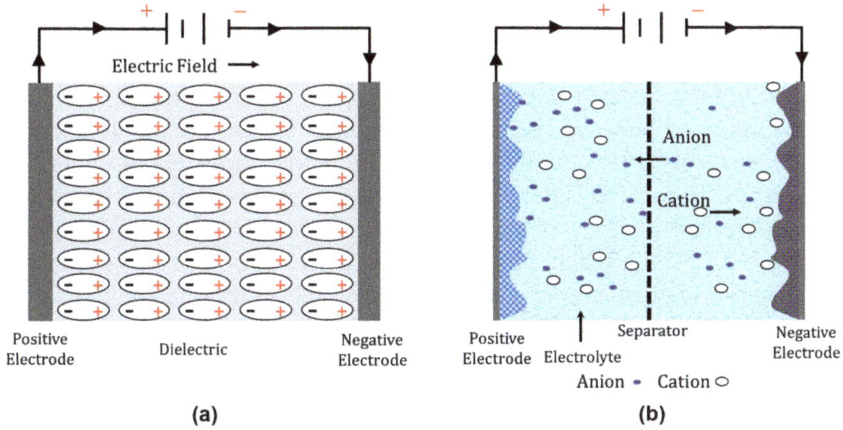

Figure 2.1. Schematics of: (a) a conventional capacitor and (b) an SC.

where ε_0 is the permittivity of vacuum, ε_r is the relative permittivity, also known as the dielectric constant, A is the effective area and d is the distance between the electrodes.

Thus, for conventional capacitors, C is directly proportional to A and inversely proportional to the distance d between the electrodes.

The performance of a capacitor is measured in terms of either energy density or power density and these performance-governing parameters are represented per unit mass or per unit volume. To calculate the power, the capacitor is connected in series with a certain load resistance, as shown in figure 2.1(a). The capacitor components such as the electrode, the dielectric material, and their interfaces also contribute to the resistance, collectively known as the equivalent series resistance (ESR).

When the value of the load resistance matches the ESR, the maximum power is delivered by the capacitor, which is given by:

$$P_{\max} = \frac{V^2}{4 \times ESR}. \quad (2.6)$$

The power loss of a capacitor in terms of the heat generated during charge and discharge is given by

$$P_{\text{loss}} = I^2 \times ESR. \quad (2.7)$$

Thus, a capacitor with a higher ESR has a greater power loss and a lower maximum power for its application.

2.3 Supercapacitors and their current position among energy storage devices

As the name suggests, SCs show promising performance when compared to conventional capacitors, although their working mechanisms are similar. The difference between them is that the electrodes of SCs have a greater surface area, they are separated by an ion-conducting separator, and the dielectric is replaced by an electrolyte (figure 2.1(b)). They offer more energy and power density than conventional capacitors. They also have a longer cycle life and a wider operating temperature range, as compared to rechargeable batteries [15–18].

When a voltage is applied across SC electrodes, it causes the accumulation of charge at the interface between the electrode and the electrolyte due to the diffusion of ions in the electrolyte toward the electrode with the opposite charge [15].

The current position of SCs lies in between those of conventional capacitors and batteries, as demonstrated by the performance matrix, known as a Ragone plot, shown in figure 2.2 [19, 20]. The plot represents the variation of power density with energy density for various energy storage devices. As represented in the figure, SCs have greater energy density and more sustainable power density than conventional capacitors. Their capacitance and power density values are also larger than those of batteries, but they have a lower energy density. Continuous efforts are being made to improve the energy density of SCs without suppressing their intrinsic power density, for example, using novel nanostructured electrode materials, combinations of

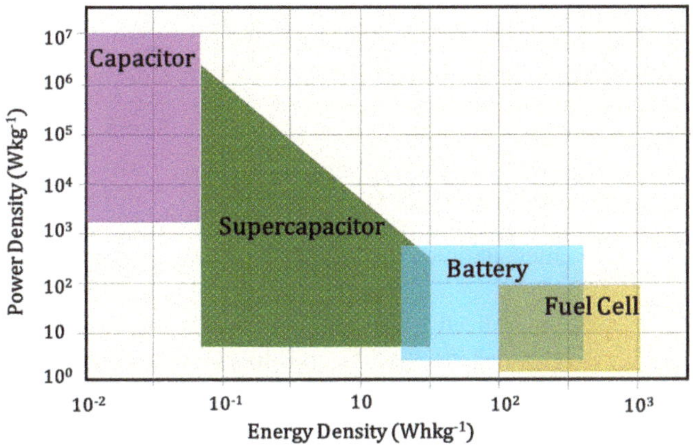

Figure 2.2. Ragone plot for various energy storage devices (specific power against specific energy).

different materials, new classes of electrolytes, and nanocomposite hybrid materials [8, 9, 17, 18]. Typically, the capacitance of an SC is governed by various parameters, as shown in equation (2.5). However, in the case of SCs, A denotes the active surface area of the electrode, ε is the dielectric constant of the electrolyte, and d is the effective thickness of the charge layers (accumulated charges) at the interfaces.

Compared to conventional capacitors and batteries, SCs have the following advantages:

- efficient and rapid energy release and absorption
- high power density
- long cycle life with little degradation
- wide window of operating temperature
- efficient operation with abundant and less toxic electrode materials
- independence from heavy metals
- the flexibility to move towards eco- and environment-friendly technologies
- larger current charge and/or discharge capacity
- low heating losses
- good reversibility
- improved safety

As well as these advantages, SCs have the disadvantages of substantial self-discharge, low energy density, and high initial cost due to the variability of prototypes and manufacturing units in the initial setup stages. Table 2.1 compares selected performance parameters of SCs and conventional capacitors.

2.4 Components of supercapacitors and their key parameters

As stated earlier, an SC has three basic components: the electrode, the electrolyte, and the separator. For a full cell assembly, two electrodes are required. The separator is sandwiched between the electrodes and then the whole assembly is immersed in liquid electrolyte to form an SC. During charging, ions within the

Table 2.1. Performance comparison between SCs and conventional capacitors.

Parameter	SC	Conventional capacitor
Charge storage mechanism	Stores charge by: • formation of a double electrostatic layer • redox reactions • intercalation	Stores charge in the form of an electric field generated in a dielectric material placed between its metal plate electrodes
Capacitance	Comparatively very high	Low
Charging/Discharging time	Typically high—milliseconds to seconds (MAX[a] ~30 s)	Typically low—picoseconds to milliseconds (MAX ~10^{-3} ms)
Charge–discharge efficiency	0.85 to 0.98	> 0.95
Operating temperature	−40 °C to 65 °C (−40 to −149°F)	−20 °C to 65 °C (− 4 to 149°F)
Energy density	Comparatively high (1 to 10 Wh kg^{-1})	Comparatively very low (< 0.1 Wh kg^{-1})
Power density	Comparatively low (MAX ~10 000 W kg^{-1})	Comparatively very high (MAX~ > 1 000 000 W kg^{-1})
Operating cell voltage	2.3 to 2.75 V	6 to 800 V
Charge–discharge cycles	10^5 to 10^6	10^6
Cost	Expensive	Comparatively very cheap
Industrial status	Manufacturing units in the growth stage, commercialized with carbon materials for high-energy applications such as hybrid electric vehicles on a small scale.	In widespread use, with well-established manufacturing units.

[a]MAX: maximum

electrolyte migrate through the separator and accumulate at the surface of the electrode material. Each of them is equally important, as their characteristics govern the overall performance of the SC. The choices of the electrolyte constituents and the active electrode materials are equally important, as they jointly affect the specific capacitance, ESR, operating voltage, stability, and the operating temperature range of the SC [15, 18–21]. The detailed requirements of each component are discussed in this section.

2.4.1 Electrode

There is a little confusion about the terminology relating to SC electrodes, because various nomenclatures have been used in previous reports; however, that does not affect our scientific understanding. Some publications call the electrodes the cathode and the anode, while others refer to the positive and negative electrodes or the working and counter electrodes. The nomenclature used for the electrodes is based on two basic fundamentals: the potential and the redox reaction. The electrode which is always at a higher positive potential is called the positive electrode, while the other, at a lower potential, is the negative electrode. Current flows from the positive to the negative electrode, which means that electrons flow from the negative to the positive electrode. However, the electrode at which the reduction reaction (the gain of electrons) happens is known as the cathode, while the electrode at which oxidation (the loss of electrons) happens is known as anode. Thus, on this basis, the cathode is positive and is known as the working electrode, while the anode is negative and is known as the counter electrode [15, 18].

2.4.1.1 Working electrode: cathode

The goal of high-energy and high-power-density SCs can be achieved with novel electrode designs and suitable choices of their chemical elements [17, 18]. As a fundamental fact, it is very clear that the capacitance of a capacitor depends on the surface area of the electrodes. However, high capacitance in the electrode does not also ensure an improvement in the other performance parameters. Thus, in order to obtain better SC performance, the electrode should possess an optimal balance of the following intrinsic characteristics:

- low-cost, abundant materials
- elements with multiple oxidation states
- high electrical conductivity
- highest possible specific surface area
- high pore density
- pore size distribution
- high thermal stability
- chemical stability
- electrochemical activity and stability
- highly hydrophilic surface, i.e. wettability
- excellent corrosion resistance
- ability to form composite materials

The materials usually used for electrodes in SCs are carbon-based materials, binary and ternary metal oxides, conductive polymers, metal chalcogenides (i.e. sulfides, selenides, and tellurides), Li-based oxide materials, and nanocomposites.

2.4.1.2 Counter electrode: the anode

If reduction occurs at the working electrode, oxidation takes place at the counter electrode. It is also used to complete the circuit of the electrochemical cell. It may or may not be made from the same material as the working electrode, based on which, the SC is classed differently. It is expected that to ease the kinetics of the electrochemical processes within the cell, the specific surface area of the counter electrode must be higher than that of the working electrode.

To study only the electrochemical properties of the working electrode, materials such as Pt, Au, carbon, and graphite are usually used as the counter electrode.

2.4.2 Electrolyte

Besides the electrodes, the electrolyte is also an important component of an SC that contributes to its long-term electrochemical and cyclic stability. The electrolyte generally has a liquid form and hence resides inside the porous structures of the electrode material and the separator.

The characteristics required in an electrolyte are:

- high ionic concentration
- high purity
- wide potential window
- low viscosity
- low resistivity
- low toxicity
- high thermal stability
- low volatility
- high dielectric constant
- high electrochemical stability
- redox potential
- low solvated ionic radius

The electrolytes used in SCs are divided into three types: (i) aqueous, (ii) organic (non-aqueous), and (iii) ionic liquids. Each of these has its merits and demerits. Figure 2.3 shows a schematic representation of the types of electrolyte with their merits and demerits.

2.4.2.1 Aqueous electrolytes

Aqueous electrolytes are the most preferable, because they have high electrical conductivity, higher ionic concentration, and a smaller ionic radius; therefore, they are able to provide high power density in SCs [8, 9, 15, 18, 20]. The preparatory processes and conditions for aqueous electrolytes can easily be optimized. However, their very narrow operating voltage window of about 1.2 V limits the energy storage

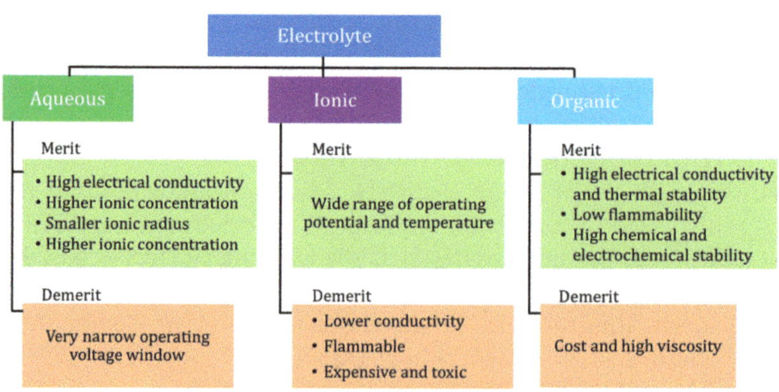

Figure 2.3. Schematic representation of the types of electrolyte with their merits and demerits.

capacity (according to equation (2.4)) of SCs made using these electrolytes. Common aqueous electrolytes include KOH, NH_4Cl, Na_2SO_4, and H_2SO_4.

2.4.2.2 Organic electrolytes

Organic electrolytes, also known as non-aqueous electrolytes, can provide an operating voltage window of about 3.5 V, which is much larger than that of aqueous electrolytes.

Organic salts, such as triethyl methyl ammonium tetrafluoroborate with solvents such as acetonitrile and propylene carbonate, are used as organic electrolytes [18, 22, 23].

Salts can dissolve in larger quantities in acetonitrile than in other solvents, but it is not environment-friendly [23, 24]. On the contrary, propylene carbonate is less toxic, environment friendly, and offers a wide range of operating potentials and temperatures. The problems with organic electrolytes are: they have comparatively lower conductivity; they are flammable, expensive, and toxic. In addition, the value of the operating voltage is quite dependent on the ppm value of the salt in the electrolyte [15, 20, 24, 25].

2.4.2.3 Ionic electrolytes

When thermal energy is provided to a salt, it melts; such a molten salt is known as an ionic liquid (IL). It remains in liquid form for a certain range of temperatures. Because of their unique features, such salts have become promising candidates for SC applications. Their features include a wide operating voltage window of about 4.5 V, high electrical conductivity, high thermal stability, low flammability, and high chemical and electrochemical stability [25–28] . One of the most important properties of an ionic liquid is that a solvation shell is not formed, as it is solvent free. The only problems are its cost and high viscosity.

Ionic electrolytes are based on either a pure ionic liquid, a eutectic ionic liquid, or a mixture of an organic solvent and ionic liquids. Ammonium salts, including asymmetric, aliphatic, and quaternary salts have been extensively investigated for use as ionic electrolytes in SCs.

It has been reported that the cyclic amines, such as aromatic pyridinium, imidazolium, and saturated piperidinium are mostly preferred due to their reasonable ionic conductivity and low viscosity compared to other ionic electrolytes [20, 27, 28].

2.4.3 The separator

A separator is an ion-permeable porous membrane used to avoid direct contact between the electrodes. Because of its high porosity, it acts as ion-transfer medium during charging and discharging. It is a very important component of SCs as far as safety is concerned. The weight, thickness, porosity, chemical composition, and morphology of a separator are important for the choice of separator in an SC [15, 18, 29].

For best performance, the separator should possess the following characteristics:

- high thermal stability
- good wettability by the electrolyte

- high porosity
- pore size and pore size distribution (submicron pore sizes are preferable)
- high tortuosity
- high dielectric strength
- minimal thickness
- chemical and electrochemical inertness
- should not be degraded by the charging and discharging processes

In general, polymeric materials are used as separators. Polypropylene film is most commonly used separator in conventional capacitors. At present, cellulose-based separators are more preferable, due to their better wettability compared to polypropylene/polyethylene. Examples of commercially available separators include [21, 29, 30]:

a. Celgard 2500 (a polypropylene-based material)
b. Solupor 14P01A (a high-molecular-weight polyethylene)
c. Gore 11367985-3 (polytetrafluoroethylene)
d. NKK TF40 (cellulose).

2.5 Performance matrixes and the measurable parameters of supercapacitors

The performance of an SC can be judged by various parameters, known as the measurable or performance-governing parameters of the SC. In detail, quantitative analysis in terms of these parameters helps us to compare SCs with other energy storage devices and to evaluate their potential for future applications. The major measurable performance parameters are the specific capacitance, energy density, power density, equivalent series resistance (ESR), self-discharge resistance, cyclic stability, and coulombic efficiency. The basics of all these are discussed in section 1.7 of chapter 1. Figure 2.4 shows the performance matrix for an SC together with the relevant measurement techniques.

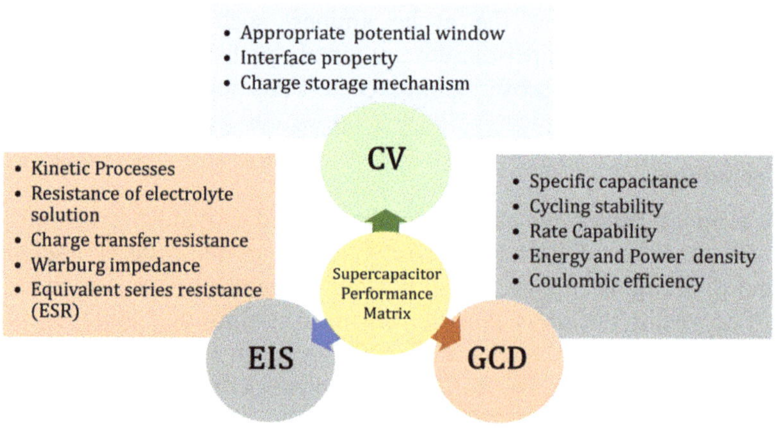

Figure 2.4. Performance matrix for an SC.

As discussed in chapter 1, the measurable parameters for electrochemical energy storage devices are represented in different units, and some of them are correlated. For a comparative study, a particular parameter should be presented using the same unit, as the representation of a parameter in a given unit is affected by its individual particular material characteristics. For example, current density can be represented either in mA cm^{-2} or mA g^{-1}. The energy density can be represented by either volumetric energy density or gravimetric energy density, with units of mAh g^{-1} and Wh kg^{-1}, respectively. Similarly, power density can be represented by volumetric energy density or gravimetric energy density, represented in W L^{-1} and W kg^{-1}, respectively.

2.5.1 Supercapacitor performance characterization

The electrochemical performance of an SC is studied using either a two- or three-electrode configuration. To test the electrochemical properties of an SC device, the two-electrode system is used, while for the characterization of a single working electrode, the three-electrode system is used. The three-electrode configuration is also known as a half cell, as there is only one working electrode, one counter, and one reference electrode; the details of this configuration are discussed in section 1.8 of chapter 1. In practice, the performance of an SC working electrode is tested in the three-electrode configuration.

The three most employed electrochemical techniques of cyclic voltammetry (CV), the galvanostatic charge–discharge (GCD) cycle, and electrochemical impedance spectroscopy (EIS) are used to investigate the electrochemical properties of SCs. Although CV and the GCD cycle can both be used to calculate the capacitance of SCs, their operating mechanisms are fundamentally different.

2.5.2 Cyclic voltammetry for supercapacitor characterization

CV is an electrochemical technique that is widely used to study the properties of the interface between the electrode and the electrolyte. The important parameters that need to be defined in CV are:

1. The starting potential
2. The switching potential
3. The final potential
4. The sweep or scan rate
5. The number of segments.

In general, the sweep rate is on the order of mV s^{-1} and can be varied from 0.1 mV s^{-1} to 1 V s^{-1} in a particular voltage window.

For an ideal capacitor, the shape of the CV curve is a perfect rectangle, as the series resistance is considered to be negligible or null. Because of series resistance, a deviation of the CV curve from its ideal rectangular shape is observed for real capacitors.

The CV plot has different shapes for different charge storage mechanisms, as illustrated in figure 1.8 in chapter 1.

2.5.3 Galvanostatic charge and discharge for supercapacitor characterization

GCD cycling is the most common method used to determine the capacitance and cycling stability of SCs. Here, each repeated loop of charging and discharging is called a cycle. The working mechanism is the same for both CV and GCD cycling; the only difference is that GCD cycling uses a current or flux of electrons to charge and discharge the capacitor instead of a driving force. The capacitance values obtained from both of these techniques are reliable but different. In GCD cycling, a constant current is provided and the voltage is monitored as a function of time; the resulting plot is known as a GCD curve. To compare CV and GCD cycling, the charging time is commonly considered. With slow charging, the capacitance values are comparable, but with fast charging times, the values are quite different.

Ideally, the GCD curve of voltage versus time should be linear, with a positive slope during charging and a negative slope during discharging. For SCs, deviation from the ideal behavior is observed in the GCD plots as well as in the CV curves, due to series resistance and a redox reaction.

In practice, due to the degradation of the electrolyte, the potential window and the slopes of the charge and discharge curves are not the same, and hence, in both cases, they may decrease. In addition, there is an instantaneous drop in voltage during the changeover from charging to discharging, due to the ESR of the system. This sudden drop in voltage is known as the IR drop.

2.5.4 Electrochemical impedance spectroscopy (EIS) for supercapacitor characterization

Electrochemical impedance spectroscopy is another technique used to characterize SCs. It is used to determine the overall impedance of an SC over a wide range of frequencies. In this method, impedance is measured as a function of frequency by applying an AC potential. The data are represented in two ways: a plot of the magnitude of the impedance ($|z|$) vs. frequency is known as a Bode plot, while a plot of the imaginary component of impedance (z'') against the real component (z') is known as a Nyquist plot.

Thus, in summary, CV, GCD cycling, and EIS are techniques based on different approaches that are used to investigate the electrochemical characteristics of SCs. In CV, the potential is linearly scanned back and forth between two extreme values at a constant rate (known as the sweep or scan rate) and the resulting current is measured. On the other hand, in GCD cycling, a constant current density is applied ($A\ g^{-1}$) and the resulting potential is measured with respect to time. EIS gives a measurement of impedance and phase shift over a wide range of frequencies. In practice, CV is used to choose an appropriate potential window. GCD cycling operates in the same potential window at different current densities to estimate the specific capacitance, energy density, power density, coulombic efficiency, and cyclic stability of SCs. The detailed analysis provided by the EIS results helps us to investigate the resistance of the electrolyte solution (R_s), the charge transfer resistance (R_{ct}), the Warburg impedance (Z_w), and the ESR.

2.6 Classification of supercapacitors

SCs are grouped into classes according to the nature of the electrode pair used and the way in which charges are generated and stored. The nature of the electrode pair depends on whether both electrodes are similar in nature or different, and the charge storage mechanism can be either faradaic or nonfaradaic. Figure 2.5 illustrates the classification of SCs and their features.

2.6.1 Classification based on structure

As discussed earlier, SCs have three major components: the electrodes, the electrolyte, and the separator. Depending upon the choice or selection of electrodes used for the fabrication of an SC, it belongs to one of three distinct families: (i) symmetric, (ii) asymmetric, or (iii) hybrid.

'Symmetric' means that both electrodes are same in terms of geometry and active electrode material; 'asymmetric' means that the electrodes are different in nature. 'Hybrid' means that the electrodes have a mixture of dissimilar materials with different charge storage mechanisms.

2.6.2 Classification based on the charge storage mechanism

The charge storage mechanisms of electrochemical energy storage devices were previously discussed in chapter 1, section 1.6. In SCs, charge is stored by the formation of an electric double layer and/or by redox reactions at the interfaces between the electrolyte and the electrode surfaces. During charging, ions accumulate at the surfaces of the electrodes, forming a layer of charges on both sides, i.e. on the electrode surface and on the electrolyte at the interface, producing capacitance at both electrodes. Here, no electrochemical reaction takes place at the electrode surface. On the other hand, if the electrode material is electrochemically active, faradaic reactions occur at the interfaces between the electrolyte and the electrode surfaces, which can directly store charges.

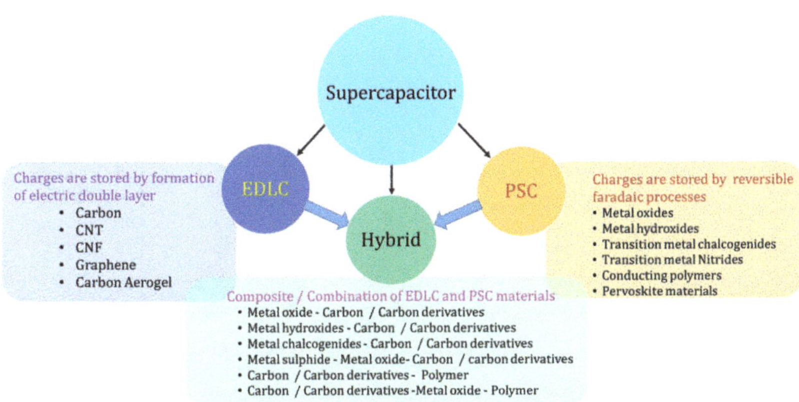

Figure 2.5. Classification of SCs and the features of each class.

Based on the charge storage mechanism and the material used for the electrodes, SCs can be grouped into three basic classes:

1. Electric double-layer SCs (EDLCs)
2. Pseudosupercapacitors (PSCs)
3. Hybrid SCs (HSCs).

In EDLCs, charges are stored electrochemically by electric double-layer formation. The materials most commonly used for electrodes are activated carbon, carbon aerogels, and graphene-based composites. In PSCs, charges are stored by faradaic reactions. The electrodes of PSCs are fabricated using either binary or ternary metal oxides, chalcogenides, or conductive polymers. In contrast, in HSCs, charges are stored using both the electric double-layer mechanism and the pseudomechanism and hence comprise hybrid electrodes with composite materials. The details of all their working mechanisms are discussed in the following sections.

2.7 Electric double-layer capacitors (EDLCs)

The working mechanism of EDLCs is nonfaradaic or electrostatic, in which charges accumulate on the surfaces of electrodes by a physical process, rather than by a chemical reaction, and form at the electrode–electrolyte surface. As no chemical change occurs during operation, EDLCs have a long cycle life along with good reversibility.

Figure 2.6 shows the typical structure of an EDLC undergoing the charging and discharging processes. It consists of two carbon electrodes, a separator, and an electrolyte. When a voltage is applied, the charges generated at the electrode produce an electric field that drives the ions in the electrolyte. The ions then migrate through the separator toward the oppositely charged electrode due to the natural tendency for attraction between unlike charges, i.e. positive ions move toward the negative electrode and negative ions move toward the positive electrode. This forms a double layer at the interface between the surface of the electrode and the

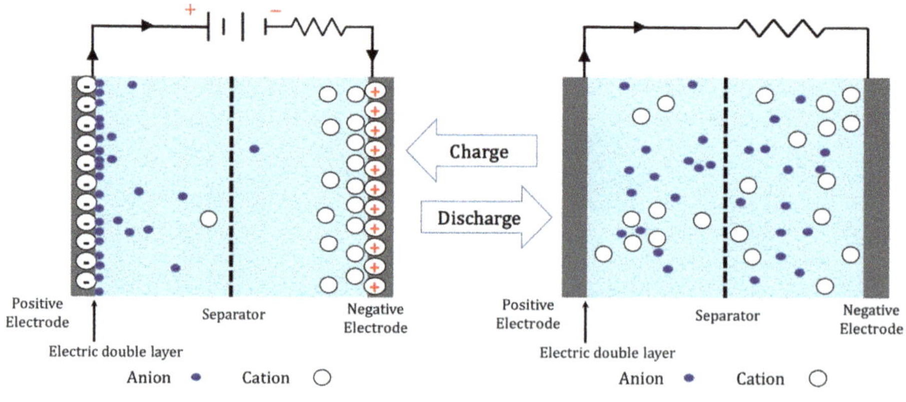

Figure 2.6. Charge–discharge process in an EDLC.

electrolyte. Thus, the total capacitance of an EDLC arises from the contributions of both electrodes.

The CV and GCD profiles for an EDLC are shown in figures 2.7(a) and (b), respectively. The resistance of EDLCs is due to the formation of the electrical double layer at the interface between the electrode and the electrolyte. This gives rise to a CV curve with a quasi-rectangular shape with rounded-off corners.

Similarly, the GCD curve is also prominently influenced by series resistance. During the changeover from charging to discharging, a deviation from linear behavior is observed, called an IR drop.

2.7.1 Working mechanism of EDLCs: the formation of an electric double layer

When a voltage is applied, polarized ions, either cations or anions, are formed at the interface, due to the arrangement of solvent molecules around the ion. Such a polarized ion is known as a solvated ion, and the shell is known as a solvated shell. Thus, the ions in an electrolyte are basically solvated ions [31, 32].

For example, consider an aqueous KOH electrolyte solution; its solvated ion structure is shown in figure 2.8. K^+ ions attract negative oxygen (negatively charged hydroxyl ions), and OH^- attracts positive hydrogen (positively charged hydronium ions) in the water molecules, respectively. This causes the formation of a shell of water molecules surrounding the ions in the electrolyte.

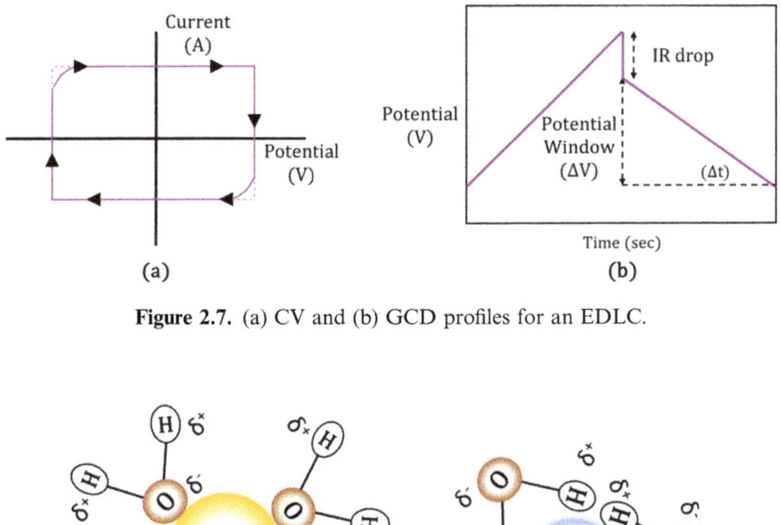

Figure 2.7. (a) CV and (b) GCD profiles for an EDLC.

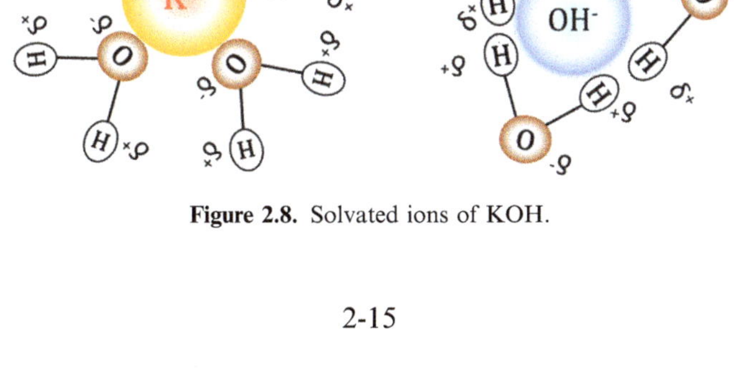

Figure 2.8. Solvated ions of KOH.

Several models have been used to explain the phenomenon of electric double-layer formation. It was first explained by Helmholtz in the nineteenth century, and hence was known as a Helmholtz layer, as shown in figure 2.9(a). According to this model, two layers of opposite charges are formed at the interface and they are separated by a distance equal to the ionic radius.

According to this model, the differential capacitance is given by

$$C_H = \frac{\varepsilon}{4\pi\delta} \tag{2.8}$$

where ε is the dielectric constant at the interface and δ is the thickness of the interface, i.e. the distance between the two layers on opposite sides.

Hypothetically, using this mechanism, the capacitance is constant and independent of the applied potential. This model became a good foundation for double-layer formation at the interfaces between two charged surfaces, even for the life sciences, but fails to explain the nature of the ion interaction and their diffusion in the electrolyte.

Further, Louis Georges Gouy and David Leonard Chapman modified Helmholtz's model by considering the diffusion mechanism of ions in the electrolyte; the resulting model is known as the Gouy and Chapman model. It is based on thermodynamics, in particular, Maxwell–Boltzmann statistics. According to this model, the capacitance is strongly influenced by the solvated ions' concentration. There is a continuous distribution of all the ions in the electrolyte because of their thermal motion [32]. They called this layer of diffused ions the diffuse layer, as shown in figure 2.9(b).

The differential capacitance according to this model is given by the following equation:

$$C_G = \frac{\varepsilon\kappa}{4\pi\delta}\cosh\frac{z}{2} \tag{2.9}$$

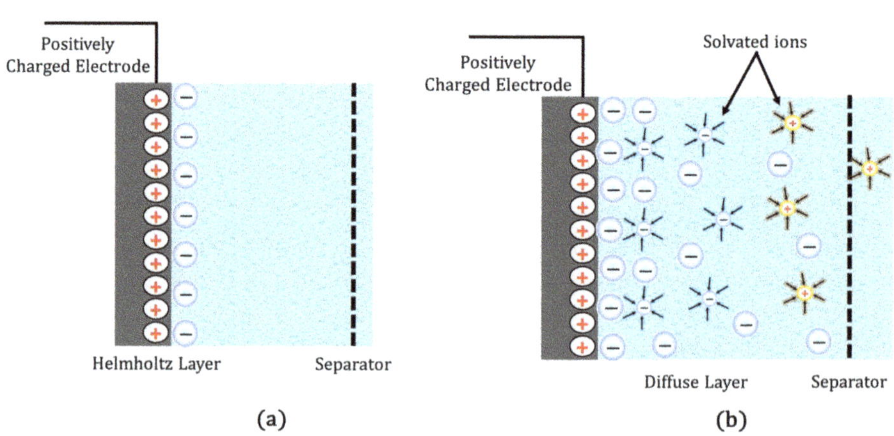

Figure 2.9. Schematic illustration of: (a) a Helmholtz layer and (b) a diffusion layer.

where z is the valence of the ions and κ is the reciprocal of the Debye–Hückel length, given by

$$\kappa = \sqrt{\frac{8\pi n e^2 z^2}{\varepsilon K T}} \tag{2.10}$$

where n is the number of ions/cm^3, K is the Boltzmann constant, and T is the absolute temperature.

The diffused charge distribution causes a concentration gradient; thus, the electric potential decreases exponentially from the electrode surface to the electrolyte, causing the variable capacitance value given by equation (2.9).

Later, in 1924, Stern introduced a new model, which was a combination of both the Helmholtz model and the Gouy–Chapman model. According to Stern, some ions adhere strongly to the electrode surface, forming a compact layer known as the Helmholtz layer, while other solvated ions form a widely distributed layer, known as the diffuse layer [32–34].

The total capacitance according to this model is a combination of both C_H and C_G defined in equations (2.8) and (2.9). It is given by

$$\frac{1}{C} = \frac{1}{C_H} + \frac{1}{C_G} \tag{2.11}$$

Nevertheless, this model is based on the assumption that the interaction between ions in the diffusion layer is purely coulombic and that the dielectric permittivity is constant throughout the layer. However, in practice, this is not the case [35].

In 1947, D. C. Grahame modified Stern's model and divided the inner compact layer into two regions. The region with ions strongly adsorbed on the electrode surface was called the inner Helmholtz plane (IHP), and the layer sandwiched between the IHP and the diffuse layer was called the outer Helmholtz plane (OHP) (sometimes referred to as the Gouy plane). The contribution of the interaction between the electrode surface charges and the solvent molecule dipoles of the solvated shell is not included in this model.

To clarify the facts about the electric double layers at the interfaces, in 1963, Bockris, Muller, and Devanathan (BMD) proposed a model, which is the current realistic model used to describe electric double-layer formation in EDLCs [33].

In the BMD model, there are three different regions of ion layer in the electrolyte. The first region, consisting of an inner compact layer at the surface of the electrode, is known as the Stern layer. It is mostly formed of adsorbed ions or molecules. These are either unsolvated anions or cations and polarized solvent molecules. The dipoles of these solvent molecules are aligned with the electrode surface due to the charges present there. These ions are referred to as specially adsorbed ions. The potential drop is strongly confined to this region, which therefore allows EDLCs to store more charges, of the order of 10^4, compared to conventional capacitors. The boundary of this layer is known as the IHP. The next region is the second region, which has an outer compact layer that consists of partially solvated cations, known as the Helmholtz layer, and its boundary is known as the OHP. The third region, with a series of diffused unevenly

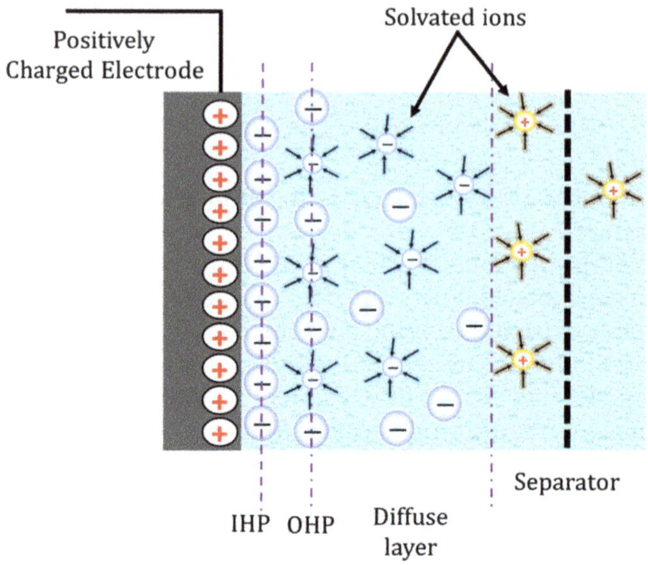

Figure 2.10. Schematic representation of the IHP, OHP, and the diffusion layer, according to the BMD model.

distributed ions, is known as the diffuse layer. The diffuse layer is formed by the thermal motion of ions in the electrolyte; it consists of positive as well as negative solvated ions [33, 36–39]. The detailed structure of the BMD model is shown in figure 2.10.

Thus, SC electrodes tend to be made from extremely porous materials, whose surface areas can be about 1000 $m^2 g^{-1}$.

In conclusion, the electric double-layer formation mechanism implies that the capacitance, i.e. the amount of charge stored, is highly influenced by the interfacial area. For greater charge accumulation, a huge electrode surface area is beneficial. An electrode material with an appropriate pore size and pore density allows rapid ionic motion and enhances the active surface area at the interface. Randomly connected macro- or micropores limit the charge storage and hence the energy and power density of EDLCs. Carbon and its derivatives, with high surface areas and regularly interconnected mesopores, are suitable for high-performance EDLCs [37–39].

2.8 Pseudosupercapacitors (PSCs)

Pseudocapacitance arises from the charge storage that occurs due to the reversible faradaic processes of charge transfer at the interface between the electrode and the electrolyte. Compared to EDLCs, which store charge electrostatically, PSCs store charge by faradaic processes (reduction–oxidation), intercalation, and electrosorption [15, 18, 40].

Pseudocapacitance can be divided into three categories according to the charge storage mechanism involved:

- underpotential deposition
- surface redox
- intercalation

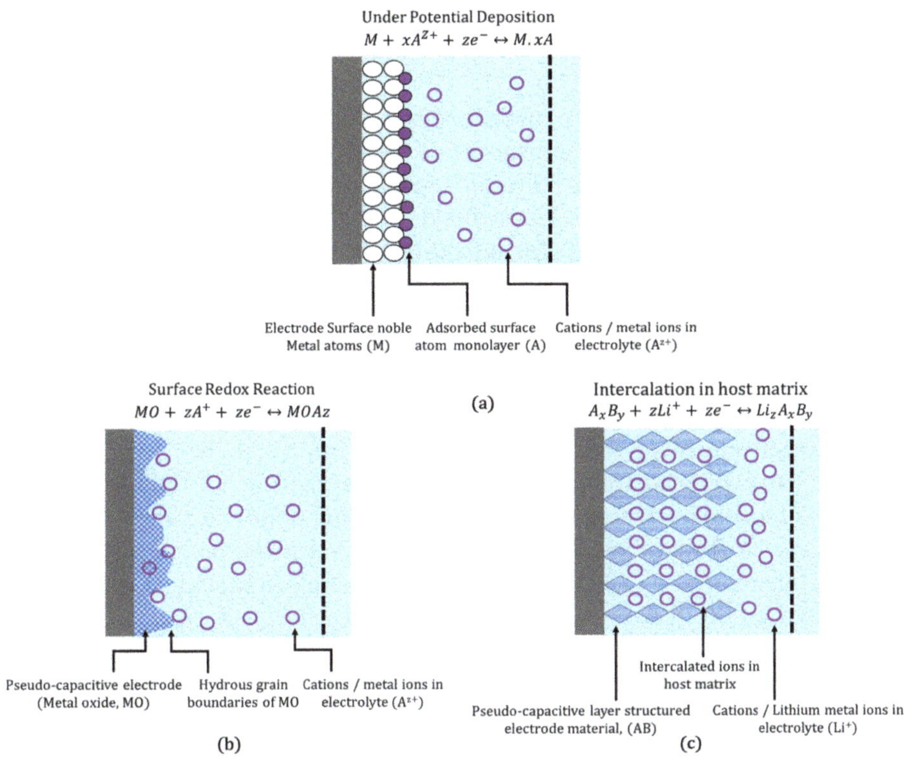

Figure 2.11. Charge storage mechanisms in PSCs: (a) underpotential deposition, (b) redox pseudocapacitance, and (c) intercalation pseudocapacitance.

Schematic representations of the different charge storage mechanisms that give rise to pseudocapacitance are shown in figure 2.11.

Pseudocapacitance due to underpotential deposition occurs when cations in the electrolyte form a monolayer on the electrode surface. The most common example is the deposition of Pb on a Au electrode at a certain potential [41].

The second main system, surface redox, is associated with faradaic charge transfer at the electrolyte–electrode interface between the electrode species and the electroactive ions of the electrolyte. In the case of transition-metal oxides, the charge transfer process occurs due to a faradaic reaction caused by the intercalation of protons (H^+) or alkaline-metal cations (A) (A can be K^+ in the case of KOH or Na^+ in the case of Na_2SO_4) from the electrolyte. This causes the formation of a reversible monolayer on the electrode surface due to the faradaic reaction, rather than electrostatics [42–45]. The reaction can be:

$$\text{For the intercalation of protons } (H^+) \Rightarrow MO + H^+ + e^- \leftrightarrow MOH \quad (2.12)$$

$$\text{For the intercalation of cations (A)} \Rightarrow MO + A^+ + e^- \leftrightarrow MOA. \quad (2.13)$$

A very recently reported and emergent type of pseudocapacitance caused by intercalation arises when electroactive ions intercalate into an electrode material with a layered structure. In this case, a fast faradaic charge transfer process takes place without any crystallographic charge in the phase of the host redox active material [45–48].

The first two pseudocapacitive mechanisms occur on the surfaces of the electrodes, while the intercalation mechanism offers more access to electrochemical active sites for ions in the electrolyte [45]. Pseudocapacitance causes peaks to appear in the CV curve and plateaus in the GCD profile during charging and discharging.

The CV and GCD profiles for all three pseudocapacitive charge storage mechanisms are shown in figure 2.12. Due to the intercalation mechanism, considerable plateaus can be observed in the GCD profile during charging and discharging, while for underpotential deposition and the redox surface mechanism, the GCD profiles are the same as for EDLCs.

When a voltage is applied to the capacitor, ions in the electrolyte start to move toward the electrode of opposite polarity. At the interface, they form an electric double layer, as explained for EDLCs. In addition, the ions in the electric double layer (some of the de-solvated ions acts as electron donors to the atoms in the electrode) cause a flow of charges, and generate a current known as a faradaic current. This generates a capacitance known as pseudocapacitance due to charge transfer caused by the redox reaction at the interface. It has been reported that the faradaic electrochemical processes extends the potential window as well as the specific capacitance of SCs.

In this case, the electrode material should be electrochemically active. The active electrode material is usually reduced to a lower oxidation state during the process of charging. This process is completely reversible, but depends upon several factors, including the electron–hopping process and proton exchange. It should be noted that the charge storage process is highly dependent on the applied voltage, hence an

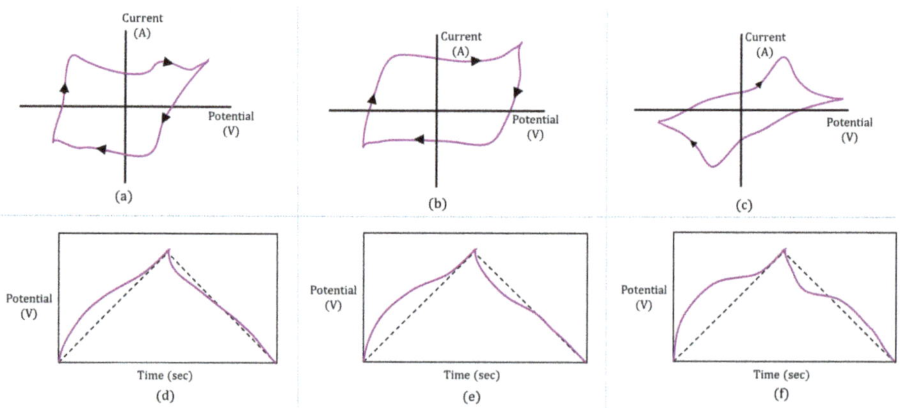

Figure 2.12. CV and GCD profiles of PSCs based on different charge storage mechanisms; CV plots of: (a) underpotential deposition, (b) the surface redox mechanism, and (c) the intercalation mechanism; GCD profiles of: (d) underpotential deposition, (e) the surface redox mechanism, and (f) the intercalation mechanism.

accurate choice of potential window is required to avoid degradation of the electrode material.

Two types of process take place at the electrodes:

1. Transfer of localized electrons
2. Transfer of delocalized electrons.

In a redox reaction, the charge transfer process results from the transfer of electrons from the ions of molecules, and is governed by the Nernst equation, which is further dependent on the oxidant and reductant species concentrations. It is also referred to as a Nernstian process. Studies have confirmed that the Nernstian process happens at a specific redox potential of the redox active material, which results in a peak-shaped cyclic voltammogram, rather than a quasi-rectangular shape as observed for EDLCs. The transferred electrons in such a process are referred to as localized valence electrons, and, in general, the process is described as a non-capacitive faradaic process. However, delocalized electron transfer, which is delocalized over wide range of energy levels in the material, is a phenomenon that is involved in pseudocapacitance, which is known as the capacitive faradaic process. This results in a quasi-rectangular shape similar to that observed for EDLCs [18, 45, 47].

The electrochemical properties of PSCs strongly depend on the electrode materials used, their surface area, and their conductivity. In recent decades, substantial efforts have been devoted to the development of novel electrode materials with promising pseudocapacitance.

The materials most commonly used for PSCs are taken from the following categories:

- metal oxides (binary and ternary)
- metal hydroxides
- transition-metal chalcogenides
- transition-metal nitrides
- conductive polymers
- perovskites
- polyoxyometalates.

2.9 Hybrid supercapacitors (HSCs)

As the name suggests, these devices work on the combined principles of EDLCs and PSCs. The merger of these mechanisms suppress the drawbacks of the individual approaches and results in improved supercapacitor performance in terms of power and energy densities. However, for this optimal combination, a mixture of both types of electrode material, i.e. carbon-based and pseudocapacitive, is required. The chemical nature, concentration, crystallographic properties, surface morphology, particle size, and the nature of the composites used in hybrid materials may affect the overall performance of an HSC. Depending upon the configuration, they have three basic classes viz. composite, asymmetric, and battery-type (the latter is the so-called supercapattery) [51–66]. Figure 2.13 demonstrates the different types of HSC with examples of the materials commonly used.

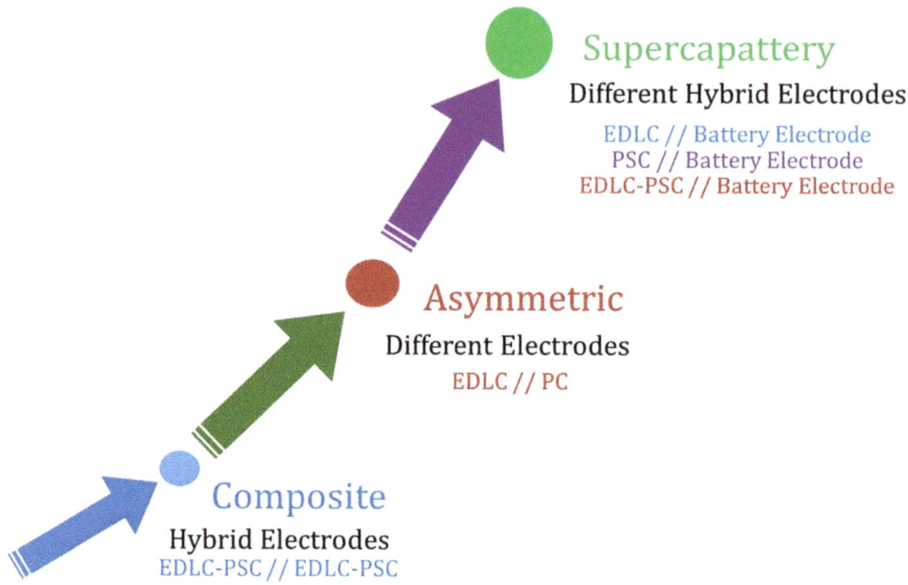

Figure 2.13. Types of hybrid supercapacitor grouped according to their designs.

2.9.1 Composite-type HSCs

In this type of HSC, the electrode is a hybrid electrode comprising both carbon and suitable pseudocapacitive materials. The carbon material offers high conductivity and a high surface area and acts as a conductive matrix between the pseudocapacitive material and the electrolyte. This combination produces capacitance due to EDLC formation as well as by faradaic reactions at the interfaces [51–55].

The presence of the composite indicates that one or more of the following modifications to the electrode structure has taken place:

- the dispersion of pseudocapacitive material in the porous carbon network
- intercalation into layered structures
- anchoring onto the surface species
- adhesion to the surface.

2.9.2 Asymmetric HSCs

In this case, instead of a composite material, an EDLC electrode is coupled with a PSC electrode; this means that one electrode is made of an EDLC material and the other is made from a pseudocapacitive material. The carbon electrode is the negative electrode and the PSC electrode is the positive electrode. This class of HSC offers a high operating voltage window and high energy and power densities, compared to ELDCs, and better cyclic stability compared to PSCs. In this case, the potential window can be widened [55–59].

2.9.3 Battery-type HSCs

This type of HSC is also known as the battery-type supercapacitor, the supercapattery, or the 'lithium-ion supercapacitor.' It consists of the coupling of a supercapacitor electrode with a battery electrode. The supercapacitor electrode can be carbon, a metal oxide, a conductive polymer, or a composite. This configuration overcomes the lower energy density limitation of supercapacitors and the lower power density limit of batteries. One studied combination consists of LTO ($Li_4Ti_5O_{12}$), $Ni(OH)_2$, with a Li metal foil as one electrode and activated carbon as the other. The battery-type electrode enhances the energy density, and the power density is maintained by the carbon electrode [60–66].

For HSCs, the degree of hybridization of the electrode materials and their structural characteristics contribute to the electrochemical performance. The correct choice of composition and a structural morphology with a mesoporous network are the key parameters for increasing the power and energy densities of HSCs, compared to those of EDLCs and PSCs.

A hybrid electrode can be a composite of any of the following forms:

- a metal oxide with carbon or carbon derivatives
- a metal hydroxide with carbon or carbon derivatives
- a metal chalcogenide with carbon or carbon derivatives
- a metal sulfide with a metal oxide and carbon or carbon derivatives
- carbon or carbon derivatives with a polymer
- carbon or carbon derivatives with a metal oxide and a polymer.

At present, massive efforts are under way to synchronize power and energy densities by the hybridization of capacitive and battery materials. The novel hybrid device known as the lithium-ion supercapacitor (LISC) can be configured in either a symmetric or an asymmetric form, depending upon the combination of electrode materials used.

2.10 Summary

The discussion in this chapter 'Perspectives on electrochemical supercapacitors: working principles and classification' sheds light on our fundamental understanding of the working mechanisms of supercapacitors in all their essential aspects and also highlights the recent progress made in different classes of supercapacitor. Supercapacitors are generally divided into three basic classes according to their charge storage mechanism: EDLCs, PSCs, and HSCs. Carbon and its derivatives are most preferable class of materials used for EDLCs, while metal oxides, metal sulfides, hydroxides, and conductive polymers are the preferred materials for PSCs. A combination of these materials is used for hybrid electrode fabrication in the case of HSCs. Recent developments in electrode materials and their hybridized forms may lead to supercapacitor technologies for future high-energy applications.

References and further reading

[1] Martinot E, Dienst C, Weiliang L and Qimin C 2007 Renewable energy futures: targets, scenarios, and pathways *Annu. Rev. Environ. Resour.* **32** 205–39

[2] Hadjipaschalis I, Poullikkas A and Efthimiou V 2009 Overview of current and future energy storage technologies for electric power applications *Renew. Sustain. Energy Rev.* **13** 1513–22

[3] Goodenough J B 2014 Electrochemical energy storage in a sustainable modern society *Energy Environ. Sci.* **7** 14–8

[4] Goodenough J B and Park K S 2013 The Li-ion rechargeable battery: a perspective *J. Am. Chem. Soc.* **135** 1167–76

[5] Yu X and Manthiram A 2018 Electrode–electrolyte interfaces in lithium-based batteries *Energy Environ. Sci.* **11** 527–43

[6] Kouchachvili L, Yaïci W and Entchev E 2018 Hybrid battery/supercapacitor energy storage system for the electric vehicles *J. Power Sources* **374** 237–48

[7] Dubal D P, Ayyad O, Ruiz V and Gomez-Romero P 2015 Hybrid energy storage: the merging of battery and supercapacitor chemistries *Chem. Soc. Rev.* **44** 1777–90

[8] Kötz R and Carlen M J E A 2000 Principles and applications of electrochemical capacitors *Electrochim. Acta* **45** 2483–98

[9] Sharma P and Bhatti T S 2010 A review on electrochemical double-layer capacitors *Energy Convers. Manage.* **51** 2901–12

[10] Kan S Y, Verwaal M and Broekhuizen H 2006 The use of battery–capacitor combinations in photovoltaic powered products *J. Power Sources* **162** 971–74

[11] Miller J R and Simon P 2008 Electrochemical capacitors for energy management *Sci. Mag.* **321** 651–52

[12] Castaings A, Lhomme W, Trigui R and Bouscayrol A 2016 Practical control schemes of a battery/supercapacitor system for electric vehicle *IET Electr. Syst. Transp.* **6** 20–6

[13] Jing W, Lai C H, Wong W S and Wong M D 2017 Dynamic power allocation of battery-supercapacitor hybrid energy storage for standalone PV microgrid applications *Sustain. Energy Technol. Assess.* **22** 55–64

[14] Zhang L, Hu X, Wang Z, Sun F, Deng J and Dorrell D G 2017 Multiobjective optimal sizing of hybrid energy storage system for electric vehicles *IEEE Trans. Veh. Technol.* **67** 1027–35

[15] Conway B E 2013 *Electrochemical Supercapacitors: Scientific Fundamentals and Technological Applications* (Berlin: Springer)

[16] Burke A 2010 Ultracapacitor technologies and application in hybrid and electric vehicles *Int. J. Energy Res.* **34** 133–51

[17] Najib S and Erdem E 2019 Current progress achieved in novel materials for supercapacitor electrodes: mini review *Nanoscale Adv.* **1** 2817–27

[18] González A, Goikolea E, Barrena J A and Mysyk R 2016 Review on supercapacitors: technologies and materials *Renew. Sustain. Energy Rev.* **58** 1189–206

[19] Holmberg S, Perebikovsky A, Kulinsky L and Madou M 2014 3-D micro and nano technologies for improvements in electrochemical power devices *Micromachines* **5** 171–203

[20] Lokhande P E, Chavan U S and Pandey A 2020 Materials and fabrication methods for electrochemical supercapacitors: overview *Electrochem. Energy Rev.* **3** 155–86

[21] Li Y, Han X, Yi T, He Y and Li X 2019 Review and prospect of $NiCo_2O_4$-based composite materials for supercapacitor electrodes *J. Energy Chem.* **31** 54–78

[22] Brandt A and Balducci A 2014 Theoretical and practical energy limitations of organic and ionic liquid-based electrolytes for high voltage electrochemical double layer capacitors *J. Power Sources* **250** 343–51

[23] Nguyen H V T, Kwak K and Lee K K 2019 1, 1-Dimethylpyrrolidinium tetrafluoroborate as novel salt for high-voltage electric double-layer capacitors *Electrochim. Acta* **299** 98–106

[24] SchŘtter C, Bothe A and Balducci A 2020 Mixtures of acetonitrile and ethyl isopropyl sulfone as electrolytes for electrochemical double layer capacitors *Electrochim. Acta* **331** 135421

[25] Zhong C, Deng Y, Hu W, Qiao J, Zhang L and Zhang J 2015 A review of electrolyte materials and compositions for electrochemical supercapacitors *Chem. Soc. Rev.* **44** 7484–539

[26] Armand M, Endres F, MacFarlane D R, Ohno H and Scrosati B 2011 Ionic-liquid materials for the electrochemical challenges of the future *Materials for Sustainable Energy: A Collection of Peer-reviewed Research and Review Articles from Nature Publishing Group* (Singapore: World Scientific) pp 129–37

[27] Watanabe M, Thomas M L, Zhang S, Ueno K, Yasuda T and Dokko K 2017 Application of ionic liquids to energy storage and conversion materials and devices *Chem. Rev.* **117** 7190–239

[28] Wong S I, Lin H, Sunarso J, Wong B T and Jia B 2020 Optimization of ionic-liquid based electrolyte concentration for high-energy density graphene supercapacitors *Appl. Mater. Today* **18** 100522

[29] Izazi A A, Lai C W, Juan J C, Phang S M, Pan G T and Yang T C 2019 Separators for supercapacitors *Supercapacitor Technol. Mater. Process. Archit.* **61** 95

[30] Jänes A, Tõnurist K, Thomberg T and Lust E 2009 Comparison of electrospun and commercially available separator materials for supercapacitors *ECS Trans.* **19** 23

[31] Schmickler W 1996 Electronic effects in the electric double layer *Chem. Rev.* **96** 3177–200

[32] Sparnaay M J 1969 Analogies between semiconductors and electrolyte solutions (in particular of surface effects due to space charges and gouy layers) *J. Chim. Phys.* **66** 87–99

[33] Oldham K B 2008 A Gouy–Chapman–Stern model of the double layer at a (metal)/(ionic liquid) interface *J. Electroanal. Chem.* **613** 131–8

[34] Matsumoto M 2018 Electrocapillarity and double layer structure *Electrical Phenomena at Interfaces* (London: Routledge)

[35] Limitation of stern model and electric double layer formation (https://web.nmsu.edu/~snsm/classes/chem435/Lab14/double_layer.html)

[36] Nakamura M, Sato N, Hoshi N and Sakata O 2011 Outer Helmholtz plane of the electrical double layer formed at the solid electrode–liquid interface *ChemPhysChem.* **12** 1430–34

[37] Yoon S, Lee J, Hyeon T and Oh S M 2000 Electric double-layer capacitor performance of a new mesoporous carbon *J. Electrochem. Soc.* **147** 2507

[38] Zhang Y, Feng H, Wu X, Wang L, Zhang A, Xia T, Dong H, Li X and Zhang L 2009 Progress of electrochemical capacitor electrode materials: a review *Int. J. Hydrogen Energy* **34** 4889–99

[39] Bleda-Martínez M J, Maciá-Agulló J A, Lozano-Castelló D, Morallon E, Cazorla-Amorós D and Linares-Solano A 2005 Role of surface chemistry on electric double layer capacitance of carbon materials *Carbon* **43** 2677–84

[40] Wang Y, Song Y and Xia Y 2016 Electrochemical capacitors: mechanism, materials, systems, characterization and applications *Chem. Soc. Rev.* **45** 5925–50

[41] Chang B Y, Ahn E and Park S M 2008 Real-time staircase cyclic voltammetry Fourier transform electrochemical impedance spectroscopic studies on underpotential deposition of lead on gold *J. Phys. Chem.* **112** 16902–9
[42] Toupin M, Brousse T and Bélanger D 2004 Charge storage mechanism of MnO_2 electrode used in aqueous electrochemical capacitor *Chem. Mater.* **16** 3184–90
[43] Rochefort D and Pont A L 2006 Pseudocapacitive behaviour of RuO2 in a proton exchange ionic liquid *Electrochem. Commun.* **8** 1539–43
[44] Jiang H, Li C, Sun T and Ma J 2012 A green and high energy density asymmetric supercapacitor based on ultrathin MnO_2 nanostructures and functional mesoporous carbon nanotube electrodes *Nanoscale* **4** 807–12
[45] Wang H, Liu J, Chen Z, Chen S, Sum T C, Lin J and Shen Z X 2017 Synergistic capacitive behavior between polyaniline and carbon black *Electrochim. Acta* **230** 236–44
[46] Conway B E, Birss V and Wojtowicz J 1997 The role and utilization of pseudocapacitance for energy storage by supercapacitors *J. Power Sources* **66** 1–14
[47] Liu J, Wang J, Xu C, Jiang H, Li C, Zhang L, Lin J and Shen Z X 2018 Advanced energy storage devices: basic principles, analytical methods, and rational materials design *Adv. Sci.* **5** 1700322
[48] Yoo H D, Li Y, Liang Y, Lan Y, Wang F and Yao Y 2016 Intercalation pseudocapacitance of exfoliated molybdenum disulfide for ultrafast energy storage *ChemNanoMat* **2** 688–91
[49] Wei M, Wei K, Ichihara M and Zhou H 2008 Nb_2O_5 nanobelts: a lithium intercalation host with large capacity and high rate capability *Electrochem. Commun.* **10** 980–83
[50] Fleischmann S, Zeiger M, Jäckel N, Krüner B, Lemkova V, Widmaier M and Presser V 2017 Tuning pseudocapacitive and battery-like lithium intercalation in vanadium dioxide/carbon onion hybrids for asymmetric supercapacitor anodes *J. Mater. Chem.* **5** 13039–51
[51] Naoi K and Simon P 2008 New materials and new configurations for advanced electrochemical capacitors *Electrochem. Soc. Interface* **17** 34
[52] Naoi K 2010 'Nanohybrid capacitor': the next generation electrochemical capacitors *Fuel Cells* **10** 825–33
[53] Haspert L C, Gillette E, Lee S B and Rubloff G W 2013 Perspective: hybrid systems combining electrostatic and electrochemical nanostructures for ultrahigh power energy storage *Energy Environ. Sci.* **6** 2578–90
[54] Lin Y H, Wei T Y, Chien H C and Lu S Y 2011 Manganese oxide/carbon aerogel composite: an outstanding supercapacitor electrode material *Adv. Energy Mater.* **1** 901–7
[55] Simon P, Brousse T and Favier F 2017 *Supercapacitors Based on Carbon or Pseudocapacitive Materials* (London: ISTE Limited)
[56] Liu Y, Hu P, Liu H, Song J, Umar A and Wu X 2019 Toward a high performance asymmetric hybrid capacitor by electrode optimization *Inorg. Chem. Front.* **6** 2824–31
[57] Liu H, Dai M, Zhao D, Wu X and Wang B 2020 Realizing superior electrochemical performance of asymmetric capacitors through tailoring electrode architectures *ACS Appl. Energy Mater.* **3** 7004–10
[58] Minakshi M, Mitchell D R, Jones R T, Pramanik N C, Jean-Fulcrand A and Garnweitner G 2020 A hybrid electrochemical energy storage device using sustainable electrode materials *ChemistrySelect* **5** 1597–606
[59] Lee J H, Shin W H, Lim S Y, Kim B G and Choi J W 2014 Modified graphite and graphene electrodes for high-performance lithium ion hybrid capacitors *Mater. Renew. Sustain. Energy* **3** 1–8

[60] Choi H S and Park C R 2014 Theoretical guidelines to designing high performance energy storage device based on hybridization of lithium-ion battery and supercapacitor *J. Power Sources* **259** 1–14

[61] Cericola D and Kötz R 2012 Hybridization of rechargeable batteries and electrochemical capacitors: principles and limits *Electrochim. Acta* **72** 1–17

[62] Yang J J, Kim Y R, Jeong M G, Yuk Y J, Kim H J and Park S G 2015 Synthesis and electrochemical characteristics of spherical $Li_4Ti_5O_{12}$/CNT composite materials for hybrid capacitors *J. Electrochem. Sci. Technol.* **6** 59–64

[63] Sun Y, Tang J, Qin F, Yuan J, Zhang K, Li J, Zhu D M and Qin L C 2017 Hybrid lithium-ion capacitors with asymmetric graphene electrodes *J. Mater. Chem.* **5** 13601–9

[64] Brisse A L, Stevens P, Toussaint G, Crosnier O and Brousse T 2018 $Ni(OH)_2$ and NiO based composites: battery type electrode materials for hybrid supercapacitor devices *Materials* **11** 1178

[65] Krause A, Kossyrev P, Oljaca M, Passerini S, Winter M and Balducci A 2011 Electrochemical double layer capacitor and lithium-ion capacitor based on carbon black *J. Power Sources* **196** 8836–42

[66] Lamb J J and Burheim O S 2021 Lithium-ion capacitors: a review of design and active materials *Energies* **14** 979

Chapter 3

Carbon-based materials for electric double-layer supercapacitors: recent advances and future perspectives

The electrochemical performance of a supercapacitor (SC) is determined by the properties and combinations of its electrode materials. Efforts have been made to tailor the physicochemical and electrochemical properties of electrode materials in order to the improve electrochemical performance of SCs. This chapter discusses the performance of electric double-layer supercapacitors (EDLCs) based on carbon and its various derivatives; it covers several classes of carbon materials, such as activated carbon, carbon fiber, carbon nanotubes, and graphene with their key features, limitations, and efforts made to improve the performance of EDLCs based on each of them.

3.1 Introduction to carbon-based electrode materials for EDLCs

The electrochemical performance of an SC is strongly influenced by the electrode materials selected; their electrical, structural, and microstructural properties play a crucial role in determining the specific capacitance, cyclic stability, and energy and power density of an SC [1, 2]. As discussed in section 2.6.2 of chapter 2, charge storage mechanisms are mainly based on surface processes, and are therefore dependent on pore size, density, and the specific surface area (SSA) of the electrode material. Larger active or contact interface areas can greatly influence the performance of EDLCs [3].

Carbon materials, with their versatile forms, have been studied extensively as electrode materials for SCs. In particular, they have proven their suitability for EDLCs by their good electrical conductivity, high SSA, and good mechanical stability. The significant properties of carbon that make it a suitable candidate for SCs are [1–5]:

- good electrical conductivity
- high electron mobility
- good chemical and thermal stability

doi:10.1088/978-0-7503-3103-6ch3

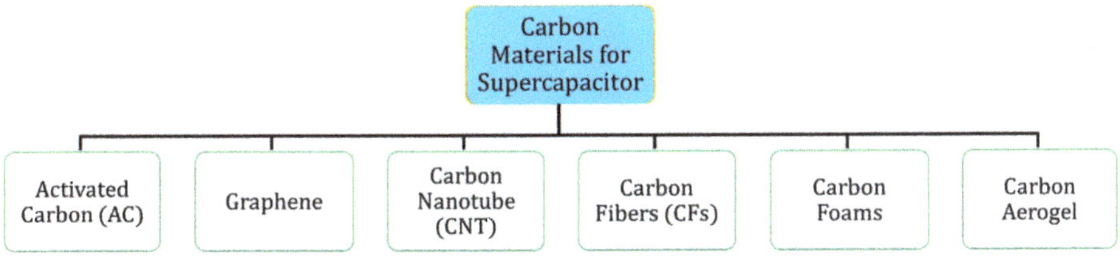

Figure 3.1. Different carbon materials used for EDLCs.

- excellent mechanical strength
- abundant resources and relatively low cost
- easy extraction from natural carbon-rich organic precursors
- large and tunable SSA
- perfect charge–discharge behavior.

The use of porous carbon, with a large accessible surface area, helps to accumulate more charges at the electrode–electrolyte interfaces. At present, different forms of carbon material, such as activated carbon (AC) [6], graphene [7], carbon nanotubes (CNTs) [8], carbon fibers [9], carbon foams [10], carbon aerogel [11], etc, are widely used as electrode materials in EDLCs. Figure 3.1 illustrates carbon-based electrode materials and their forms reported for supercapacitor applications.

3.2 Activated carbon (AC)

AC is the most preferable material for SC fabrication, even in industry, because of its abundant resources, porous nature, and moderate cost. From the literature, it can be observed that research into AC materials is mainly focused on optimizing their pore size, shape, density, regularity, and pore size distribution, along with efforts to enhance their electrical properties and SSA. Significant access to the pores in the electrode material by the electrolyte is important for improved EDLC performance. The factors that influence the electrochemical performance of AC-based EDLCs are [12, 13]:

- electrical conductivity
- SSA
- pore size and shape
- pore size distribution
- network connection between pores
- functionality of the surface.

Activated carbon is produced via two steps, carbonization and activation. During carbonization, the volatile content that pre-exists in the carbon precursor is reduced by pyrolysis at temperatures in the range of about 300 °C–900 °C, while the activation process is performed to improve the SSA by creating a porous network. The methodology used for activation can produce a complex distribution of pores, in terms of their size and shape, which is beneficial for the electrochemical performance of EDLCs [13, 14]. Figure 3.2 illustrates different approaches for

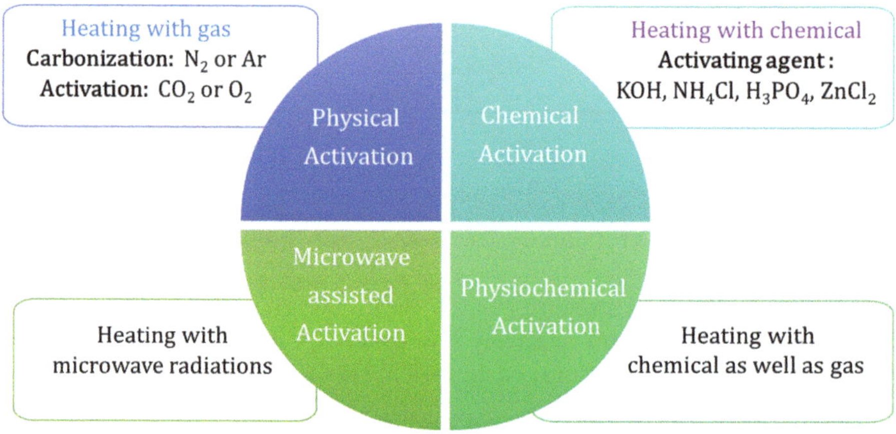

Figure 3.2. Methods of carbon activation.

activating carbon, such as physical, chemical, physicochemical, and microwave-assisted activation.

3.2.1 Physical activation

In physical activation, carbon is processed by heating the raw material in the absence of atmospheric air at very high temperatures, usually in the range of 700 °C–1200 °C. During physical activation, carbonization is followed by activation; therefore, it is known as a two-stage process. The carbonization is performed in the presence of an inert gas such as nitrogen (N_2) or argon (Ar), subsequently followed by activation using an oxidizing gas such as carbon dioxide (CO_2) or oxygen (O_2). Since it produces no secondary waste materials, activation is classed as a green approach to producing porous AC materials. The factors affecting physical activation are the gas flow rate, temperature, heating rate, reaction rate of pyrolysis, carbonization temperature, and time. One of the major disadvantages of physical activation is that the activation temperature and processing time vary according to the carbon precursor used. It also has a low carbon yield and results in a low SSA, compared to that resulting from chemical activation.

3.2.2 Chemical activation

Chemical activation is a single-stage process used for the production of high yields of AC. The carbonization and activation processes are performed simultaneously by heating the carbon source in the presence of an activating agent at a temperature from about 400 °C to 700 °C. In general, activating agents are classified into four categories, namely, alkaline, acidic, neutral, and self-activating agents [15]. As a rule of thumb, acidic and self-activating agents form macropores, alkaline agents generate micropores and neutral agents produce mesopores. The most common and recently reported activating agents are KOH [16], NaOH [17], NH_4Cl [18], H_3PO_4 [19], $ZnCl_2$ [20], $FeCl_3$ [20], H_2SO_4 [21], K_2CO_3 [22], Na_2CO_3 [22], $KMnO_4$ [23], $NaNH_2$ [24], etc.

Table 3.1. Comparison of physical and chemical activation processes.

Parameter	Physical activation	Chemical activation
Methodology	Two-stage process: carbonization is followed by activation	Single-stage process: carbonization and activation are performed simultaneously
Activating agent	**Carbonization:** Nitrogen (N_2) or argon (Ar), **Activation:** Carbon dioxide (CO_2) or oxygen (O_2).	KOH, NaOH, NH_4Cl, H_3PO_4, $ZnCl_2$, $FeCl_3$, H_2SO_4, K_2CO_3, Na_2CO_3, $KMnO_4$, $NaNH_2$, etc.
Factors affecting activation	Gas flow rate, temperature and heating rate, carbonization temperature and time, reaction rate of pyrolysis	Activating agent, heating temperature, carbon precursor, ratio of activating agent to carbon precursor
Yield	Low carbon yield and low SSA	High carbon yield and high SSA

The factors affecting chemical activation include the heating temperature, activating agent, and carbon precursor and the ratio of activating agent to carbon precursor used during the activation process. Compared to physical activation, chemical activation has a low processing temperature, a high carbon yield, and a high SSA. The pore size and distribution can easily be controlled by varying the type and ratio of the activating agent to the carbon precursor [15–24]. To reduce the heating time, chemical activation can be assisted by microwave radiation, which is then known as microwave-assisted activation. In microwave-assisted activation, pyrolysis is performed using microwave radiation with sufficient energy. Table 3.1 shows a comparison of physical and chemical activation processes.

3.2.3 Physicochemical activation

Recently, trials of physicochemical activation have taken place, which is a combination of both the physical and chemical activation procedures. The motive for the trials was to control the surface texture and porous nature of AC. During physicochemical activation, carbonization and activation are performed using an activating agent in an oxidizing gas atmosphere. Although it seems to be beneficial at a small or laboratory scale, its complex procedure and large energy consumption obstruct its use for large-scale production.

3.2.4 Performance of AC-based EDLCs

Several trials have been performed with the aim of improving the performance of EDLCs with AC by enhancing the SSA and optimizing the pore size using plasma surface treatments, alkaline treatments, CO_2 activation, and heat treatments [12, 25, 26]. These approaches effectively create surface defects and influence the porous structure of the carbon material, and hence enhance the SSA. Ideally, one might expect an increase in the specific capacitance from an increased SSA; however, this is sometimes impossible. Several studies have shown that carbon electrodes with smaller SSAs have a higher specific capacitance than electrodes with larger surface areas [13, 27]. This is because a more appropriate pore size distribution exists in the former. According to their sizes, pores roughly fall into three classes: micropores, macropores, and mesopores. Micropores have a size smaller than 2 nm, macropores have a size greater than 50 nm, while the mesopore size is in the range of 2 to 50 nm. The porous structure should have adequate accessibility for ions, i.e. solvated ions in the electrolyte. It should be noted that the mobility of ions in a porous network structure is different from that of the bulk [13, 26, 27].

Even though conventional carbon materials have good conductivity, large surface areas, and extensive porous networks, they still exhibit poor performance in EDLCs. This is because of the random distributions and connections of the pores in their structures [28].

In the case of carbon electrodes with micropores, solvated ions may not reach every micropore, hence, the whole SSA does not contribute to charge storage due to the inaccessibility of some pores, leading to poor EDLC performance [29]. Thus, an increase in the density of micropores in the material can increase the SSA but not necessarily the specific capacitance. Consequently, suitable matching of the pore size, its distribution in the electrode, and that of the solvated ions in the electrolyte can enhance AC-based EDLC performance. Mesoporous carbon offers a high SSA and a fast ion transport pathway, which contributes to the high power capability of SCs [28, 30]. As an example, Yamada, H *et al* [31] have synthesized porous carbon with meso-/macropores (8–80 nm) and micropores, which showed an enhanced specific capacitance of 350 F g^{-1} at 1 mV s^{-1} in 1 M H_2SO_4 electrolyte, as an effect of the balance between micro- and mesopores.

However, the mesopore size and distribution play a crucial role in kinetic processes and hence influence the electrochemical performance of EDLCs. An even distribution of pores can help to enhance ion mobility by reducing the ion transport length in the pore structure [32–39].

Furthermore, the performance of EDLCs is not same with aqueous and organic electrolytes for the same electrode, with either micro- or mesopores. This is because the size of the solvated ions is not same for both electrolytes. According to a previous report, an EDLC with AC showed a higher specific capacitance value of about 300 F g^{-1} in an aqueous electrolyte than in an organic solution (about 150 F g^{-1}), because the ion size is larger for the organic electrolyte [33]. Thus, the micropores play an important role for charge storage, and hence, the formation of the electrical double layer; however, mesopores are crucial for charge or ion transport. An effective population distribution of mesopores and micropores in the electrode material that suits the electrolyte ensures the efficient electrochemical performance of EDLCs.

According to previous reports, a pore size of 0.8 nm can be suitable in the case of organic electrolytes, while pore sizes of 0.4 or 0.7 nm can be useful for aqueous electrolytes [12]. A carbon electrode with a hierarchical pore structure containing micropores for charge storage, mesopores for ion transport, and macropores for the infiltration of the electrolyte would be an ideal choice for an EDLC.

For instance, Xia *et al* [34] prepared activated mesoporous carbon with CO_2 at 950 °C. The CO_2 treatment caused a remarkable improvement in the capacitive performance. The specific capacitance improved from 115 to 223 F g^{-1} in 6 M KOH electrolyte and showed a retention stability of about 73%. The observed improvement was attributed to the hierarchical pore structure, the balanced mesopore to micropore ratio, the high SSA of about 2749 m^2 g^{-1}, and the large pore volume of about 2.09 cm^3 g^{-1}. Raymundo-Piñero *et al* [12] reported the synthesis of activated carbons by the heat treatment of bituminous coal at temperatures in the range of 520 °C–1000 °C and subsequent activation by KOH at 700 °C or 800 °C. The results showed an increase in the surface area of the activated carbon from 800 to 3000 m^2 g^{-1}, depending on the temperature at which carbonization took place. Electrochemical measurements using KOH and H_2SO_4 electrolytes showed specific capacitance values of 124–286 and 160–310 F g^{-1}, respectively. Li *et al* [35] synthesized mesoporous activated carbon using lignite at a temperature of 650 °C. A specific capacitance of 207.5 F g^{-1} at a current density of 0.5 A g^{-1} in 6 M KOH electrolyte with a retention of about 91% at a current density of 55 A g^{-1} over 3000 cycles was obtained as a result of a SSA of 1024 m^2 g^{-1}.

Redondo *et al* [36] reported the effect of pore texture on the performance of activated carbon supercapacitor electrodes and achieved a specific capacitance of 260 F g^{-1} at 0.5 A g^{-1}. The authors argued that an EDLC with a moderate KOH/C ratio and comparable sizes of pores and solvated ions in its electrolyte can provide effective high-rate operation in EDLCs. Pores whose sizes are closer to the sizes of the solvated ions contribute to enhancing the low-rate specific capacitance; conversely, excessive solvation in large micropores decreases capacitance retention at high rates. Also, pores that are wider than the solvated ion size allow freer ion transport at high rates. The authors concluded that pores about 25% wider than the size of the solvated electrolyte ions are conducive to better EDLC performance. Luo *et al* [37] reported the electrochemical performance of an EDLC with a 3D sponge-like nanoporous carbon electrode. Their high-temperature activation process caused an enhancement in the SSA of about ~ 1018 m^2 g^{-1}, which resulted in a high capacitance of 296 F g^{-1} at 0.7 A g^{-1} and a good capacity retention of 95.1% at 3 A g^{-1} in 6 M KOH electrolyte.

Furthermore, the doping of carbon material with heteroatoms also enhanced the electrochemical performance of EDLCs. Such doping can be done after the activation process via a post treatment with a heteroatom-containing precursor. In particular, nitrogen doping in carbon created much better performance due to improvements in the electrical conductivity, surface wettability, and mesoporous network. Heteroatoms were also found to contribute specific capacitance by enhancing the surface polarity and electron donor affinity of carbon [40–44].

For instance, Wang *et al* [40] prepared nitrogen-doped 2D porous carbon sheets from clover biomass, which showed a high specific capacitance of 436 F g^{-1} at 1 A g^{-1} and an excellent rate capacity, with a remaining capacitance of 290 F g^{-1} at 50 A g^{-1} in 1 M H_2SO_4 electrolyte. The assembled symmetric supercapacitor showed good cyclic stability, with a retention of 99.4% at 5 A g^{-1} after 30 000 cycles. The enhanced performance was attributed to improved electrical conductivity and interconnected hierarchical pore structures, which provided more electrochemical active sites and fast ion diffusion channels.

Li *et al* [41] reported the electrochemical performance of nitrogen-doped hierarchical porous carbons in different electrolytes. They prepared nitrogen-doped hierarchical porous carbons at 750 °C using CH_3COOK as an activating agent. A sample prepared with a 15 wt% CH_3COOK aqueous solution showed specific capacitances of about 284 and 355 F g^{-1} at a current density of 0.1 A g^{-1} in 6 M KOH and 1 M H_2SO_4 electrolytes, respectively. The sample also exhibited a capacitance retention of about 80% at the high current density of 10 A g^{-1} in a H_2SO_4 electrolyte. The energy density at a specific power density was also calculated, which had values of about 9.9 and 12 Wh kg^{-1} at a power density of 50 W kg^{-1} in KOH and H_2SO_4 electrolytes, respectively. According to the authors, the excellent electrochemical performance was attributed to a large SSA, nitrogen functionalities, and a hierarchical pore structure.

The higher performance of H_2SO_4 as an electrolyte, compared to that of KOH, is attributed to its easy access to micropores, smaller ion size, the improved wettability of the carbon electrode, and the reaction between heteroatoms and H^+[41]. In recent years, remarkable efforts have been made to produce mesoporous activated carbon from unconventional materials, such as biomass or biowaste. Such approaches have provided opportunities to move towards sustainable development by the effective use of biowaste for the synthesis of porous carbons. Several sources of biomass, such as plants, vegetables, plant leaves, food waste, coconut shell, wood, rice husk, etc, have been rigorously investigated for the production of highly porous carbon materials and their use in EDLCs [45–60].

For example, multihierarchical porous carbon prepared from a typical food waste, crab shell, exhibited specific capacitances of 322.5 and 223.4 F g^{-1} at current densities of 1 and 10 A g^{-1}, respectively [50]. In another novel work, hierarchical porous active carbon was prepared from fallen leaves using K_2CO_3, with KOH as an activating agent. An EDLC fabricated using this material as the electrode showed a high specific capacitance of up to 242 F g^{-1} at 0.3 A g^{-1} in 6 M KOH electrolyte. The study reported that the hierarchical porous structures and SSA were associated with the mass ratio of K_2CO_3 and KOH [51].

Yin *et al* [52] prepared activated carbon from coconut fibers with a multitubular hollow structure using KOH as the activating agent, which exhibited a specific capacitance of 266 F g^{-1} at 0.1 A g^{-1} and good cyclic stability, with a retention of about 76% of its capacitance at 100 A g^{-1} in 6 M KOH electrolyte. They found that the mass ratio of KOH and coconut fiber had an enormous effect on the specific surface area and pore volume. A sample prepared with a mass ratio 4:1 of KOH to coconut fiber showed a SSA of about 2898 m^2 g^{-1} with a pore volume about 1.59 cm^3 g^{-1}. Misnon *et al* [53] prepared carbon from an oil palm kernel shell using chemical and physical activation processes. A sample prepared by chemical activation exhibited a specific capacitance of 210 F g^{-1} at 0.5 A g^{-1} in 1 M KOH electrolyte and good cyclic stability, with a retention of 95%–97% of the capacitance after 1000 cycles. The authors observed that the value of specific capacitance of a physically activated sample was about 50% less (123 F g^{-1}) than that of a chemically activated one. Figure 3.3 displays the cyclic voltammetry (CV) curves of chemically activated carbon (AC-C) (figure 3.3(a)) and physically activated carbon (AC-P)

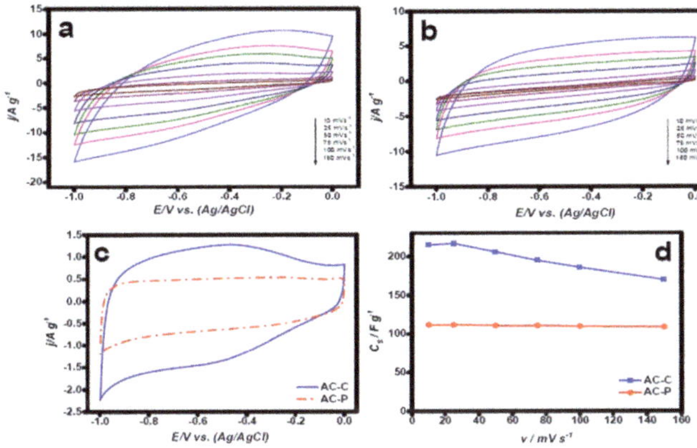

Figure 3.3. CV curves of: (a) chemically activated carbon (AC-C) and (b) physically activated carbon (AC-P) in 1 M KOH electrolyte at scan rates from 10 to 150 mV s^{-1}; (c) AC-C and AC-P at a scan rate 10 mV s^{-1} and (d) the dependence of C_S on the scan rate of CV (calculated from the CV results for AC-C and AC-P). Reprinted from [53], Copyright (2015), with permission from Elsevier.

(figure 3.3(b)) in 1 M KOH electrolyte at different scan rates from 10 to 150 mV s^{-1} and for a potential range of −1.0 to 0.0 V. The similar quasi-rectangular rectangular curves are characteristic of the EDLC charge storage mechanism.

To get a clear idea of the effects of the activation process on charge accumulation, CV curves measured at a scan rate of 10 mV s^{-1} for both AC-C and AC-P are compared and shown in figure 3.3(c). It can be observed that the shape of the CV curve for the sample prepared by physical activation (AC-P) has a more rectangular shape than that of the sample prepared by chemical activation (AC-C), indicating the pure EDLC charge storage mechanism of AC-P. However, AC-C can deliver more capacitance than AC-P. The variation of specific capacitance (C_s) calculated from the cathodic part of the CV data for different scan rates is shown in figure 3.3(d). It can be observed that C_s is independent of the scan rate in the case of AC-P, while it decreases with the scan rate in the case of AC-C. Such a change of C_s with scan rate is attributed to surface functional groups that may be responsible for faradaic reactions at the electrode surface. The cyclic performance of both samples was measured using the galvanostatic charging and discharging (GCD) technique at 0.5 to 10 A g^{-1}. Figures 3.4(a) and (b) show the GCD curves for both AC-C and AC-P in 1 M KOH electrolyte. Both samples showed ideal charging and discharging behaviour, as the shapes of the GCD curves are nearly triangular. Figure 3.4(c) compares the GCD curves of both AC-C and AC-P in 1 M KOH at a current density of 1 A g^{-1}. The AC-C electrode showed increased charging and discharging times compared to the AC-P electrode, due to the presence of functional groups at the surface, which led to an increase in the specific capacitance. Figure 3.4(d) shows the variation of C_s for AC-C and AC-P at various current densities ranging from 0.5 to 10 A g^{-1}. With an increase in the current density from 0.5 to 10 A g^{-1}, C_s can be seen to be change from 80 to 210 F g^{-1} for AC-C and from 41 to 123 F g^{-1} for AC-P, respectively. Wang *et al* [54] reported the electrochemical

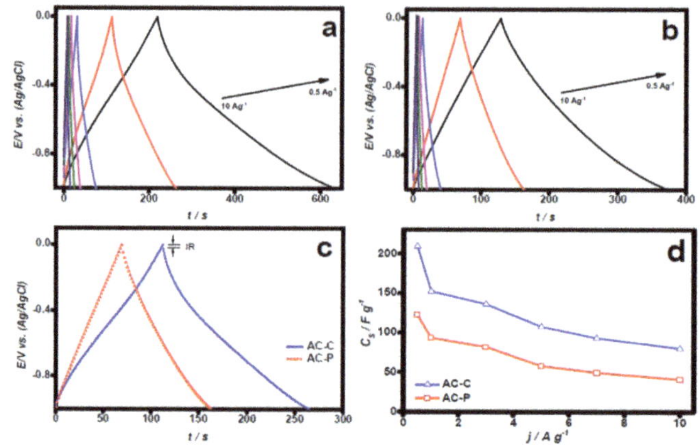

Figure 3.4. GCD curves of: (a) chemically activated carbon (AC-C) and (b) physically activated carbon (AC-P) in 1 M KOH electrolyte for a potential window of −1.0 to 0.0 V; (c) AC-C and AC-P at 1 A g^{-1} and (d) C_S at various current densities (calculated from the GCD curves for AC-C and AC-P). Reprinted from [53], Copyright (2015), with permission from Elsevier.

performance of hierarchical porous carbon (HPC) sheets directly prepared from cost-effective non-toxic biomass (cornstalks) in an air atmosphere. The HPC sheets exhibited a high SSA of 1588 m^2 g^{-1}, a specific capacitance of 407 F g^{-1} at 1 A g^{-1}, and good cyclic stability with a retention of 92.6% of the capacitance after 20 000 cycles at 5 A g^{-1}. According to the authors, the enhanced electrochemical performance was attributed to rich O doping, which introduced pseudocapacitance and hierarchical pores, facilitating electrolyte diffusion and ion transfer.

Thus, the crucial factors in developing AC-based EDLCs with superior performance are related to: the electrical conductivity, the presence of a hierarchical porous network, the accessibility of pores by electrolyte ions, the ionic conductivity, the viscosity, and the electrochemical stability of the electrolyte. Table 3.2 summarizes the electrochemical performance of EDLCs that incorporate activated carbon prepared using different sources and methods.

In conclusion, the performance of AC-based EDLCs can be improved by:

- a suitable combinations of pore size and distribution
- optimization of the pore size according to the electrolyte used
- the use of non-aqueous electrolytes for a better potential window
- surface modification
- N doping
- optimizing the temperature and atmosphere of the carbonization process
- tuning the precursor-to-activating-agent mass ratio.

3.3 Performance of EDLCs based on carbon nanotubes (CNTs)

CNTs are categorized as 1D carbon nanostructured materials with a high aspect ratio. They also possess good electrical conduction and transport properties, which facilitate a continuous network for charge transport at the interfaces [8, 61, 62]. Based on the number of carbon sheet layers utilized during formation, CNTs can

Table 3.2. Electrochemical performance of EDLCs based on activated carbon prepared in different forms.

Working electrode (SSA (m^2 g^{-1}))/(pore volume, V_p (cm^3 g^{-1}))	Electrolyte	Potential window (V)	Specific capacitance (F g^{-1}) at (@) current density (A g^{-1})/scan rate (mV s^{-1})	Power density (W kg^{-1})	Energy density (Wh kg^{-1})	Retention % (cycles) at (@) current density (A g^{-1})/scan rate (mV s^{-1})	Ref.
Porous C (SSA~2749) /(V_p~2.09)	6 M KOH	−1.0 to 0.1	223/2 and 162/50			Capacitance retention of 73%/50	[34]
AC	6 M KOH		286/2				[12]
Ordered porous carbons with meso-/macropores	1 M H_2SO_4	0.3 to 0.6	350/1			97% (200)	[31]
Microporous carbon	6 M KOH		260@0.5				[36]
AC (SSA~2934.60) /(V_p~1.33)	0.5 M K_2SO_4	−1 to 0	198.15 @0.1 and 135.19 @2	3487	82.93	~99% (2000)	[38]
3D sponge-like nanoporous carbon (SSA~1018)	6 M KOH	−1 to 0	296@0.7			95.1% (5000)	[37]
Honeycomb-like HPC (SSA~2494)/(V_p~1.51)	6 M KOH	−0.3 to −0.1	210.2@5	251.0	52.6	96.3% (10 000)	[39]
N-doped 2D porous carbon sheets	1 M H_2SO_4	0 to 1.0	436@1		58.4	99.4% (30 000)	[40]
N-doped HPC sheets	1 M H_2SO_4	0 to 1.0	355@0.1	50	12.3		[41]
	6 M KOH		284@0.1	50	9.9		
Hierarchical porous N-doped carbon submicrospheres (SSA~3155) /(V_p~2.2)	6 M KOH	−0.8 to 0.2	455@0.5		93.6	95% (5000)	[42]

N/S co-doped graphene-like carbon nanosheets	6 M KOH	−1 to 0	262.43 @1	12 312	22.26	96.28% (10 000)	[43]
N/S dual-doped HPC	1 M Na_2SO_4	−1 to 0	320@0.5		42.5	90.3% (5000)	[44]
AC from coconut fibers	6 M KOH	0 to 1.0	266@0.1	8224	53	Capacitance retention 76% @100	[52]
Porous AC from oil palm kernel shells	1 M KOH	−1 to 0	210@0.5			95%–97% (1000)	[53]
Carbon sheets from biomass (SSA~1588)	1 M H_2SO_4	0 to 1	407@1413@0.5	123.9	14.35	92.6% (20 000)	[54]
Porous carbon from rice husks	6 M KOH	0 to 1.0	110@0.1			78% (5000)	[55]
Nanoporous carbon from java kapok tree shells	6 M KOH	0 to 1	169/20–200	1900	12.5	97% (10 000)	[56]
Highly porous willow wood-derived AC	6 M KOH	−1 to 0	394@1	10 000	23	94% (5000)	[57]
Rod-like porous carbon from tea waste	6 M KOH	−1 to 0	332.1@1			97.8% (10 000)	[58]
HPC from wood sawdust	6 M KOH	−1 to 0	286@1	9213	5.63	99.8% (10 000)	[59]
N-doped microporous carbon (SSA~1186) /(V_p~2.2)	6 M KOH	−1 to 0	245@0.5			93.1% (5000)	[60]

typically be classified into two groups, namely, single-walled carbon nanotubes (SWCNTs) and multiwall carbon nanotubes (MWCNTs). Both SWCNTs and MWCNTs have been extensively studied to assess their suitability for use as energy storage electrode materials due to their unique features, such as their good electrical conduction, low mass per unit volume, high mechanical strength and flexibility, high thermal conductivity, and optimal electrochemical as well as chemical stability [63, 64]. These unique features of CNTs arise from their hexagonal lattice structure of graphene with sp^2 hybridized covalently bonded carbon atoms [65]. CNT electrodes have a lower ESR than that of AC, because their porous structure with a continuous network provides an easy pathway for electrolyte ions to move freely. However, EDLCs fabricated using CNTs have a lower energy density than those fabricated using ACs due to their relatively smaller surface area of about 500 m^2 g^{-1} [61, 63, 64].

Several attempts have been been made to enhance the porosity and SSA of CNTs, either by surface activation (of the walls and/or tips of CNTs) or by surface modification with an active material. Nevertheless, an appropriate balance between the SSA and the electrical conductivity must be realized in order to obtain good electrochemical performance from the device. The performance of EDLCs fabricated using SWCNTs is strongly affected by the size of the electrolyte ions and the ease with which they can access the electrode materials [66]. The estimated theoretical SSA for SWCNTs has a value of about 1315 m^2 g^{-1}, while that for MWCNTs is comparatively smaller [67, 68]. However, MWCNTs have more mesopores than SWCNTs, which allows ions to easily penetrate the electrode–electrolyte interface. However, it has also been observed that the mechanical stability of electrodes with CNTs decreases due to a significant volume change with repetitive charge–discharge cycles [67].

A supercapacitor based on SWCNTs has shown a maximum specific capacitance of 180 F g^{-1}, a high power density of 20 kW kg^{-1}, and an energy density of ~7 Wh kg^{-1} in a KOH electrolyte [63]. However, an EDLC based on MWCNTs with a surface area of 430 m^2 g^{-1} showed a specific capacitance of 102 F g^{-1}, a power density of about 8 kW kg^{-1}, and an energy density of 1 Wh kg^{-1} in an H_2SO_4 acidic electrolyte [68].

Moreover, the hierarchical structures of CNTs may lead to better electrochemical performance. Electrodes with aligned CNTs appear to be more beneficial in terms of power density enhancement than randomly entangled CNTs, because, in the latter case, the irregular pore structures and large degree of entanglement hamper the ionic transportation rate [69]. For instance, Kim *et al* [70] fabricated high-performance supercapacitors using vertically aligned single-walled carbon nanotube arrays, which exhibited a specific capacitance of 75 F g^{-1}, a power density of about 987 kW kg^{-1}, and an energy density of about 27 Wh kg^{-1} in an ionic liquid electrolyte (1 M tetra ethyl ammonium tetrafluoroborate in propylene carbonate). According to the result analysis, the high power and energy density could be attributed to the fast ion transport due to the oriented carbon nanotubes and the wide potential window offered by the ionic liquid. A study carried out by Shah *et al* [71] on EDLC-based aligned carbon nanotubes directly attached to conductive substrates reported that the electrochemical performance of the device was significantly influenced by the length of the CNTs. The specific capacitance, energy, and power density increased with a decrease in nanotube length from 10.75 to 21.57 F g^{-1}, 2.3 to 5.4 Wh kg^{-1}, and 19.6 to 35.4 kW kg^{-1}, respectively [71].

According to a study carried out by Li *et al* [66], the ion size and the electrode–electrolyte accessibility are also prominent factors affecting the electrochemical performance of CNT-based SCs. According to the authors of this study, when the pore size became comparable to the size of the solvated ions, a greater SSA was accessible by the electrolyte, and hence, there was an easy flow of ions across the electrode–electrolyte interface.

The electrochemical performance of CNT-based EDLCs can also be increased by improving the ease of ion transportation, optimizing the pore size and distribution, improving the conductivity, and by the addition of a pseudocapacitive charge storage mechanism. Several approaches have been reported for this purpose, based on functionalization [72–74], chemical activation [64, 75, 76], and heteroatom doping in CNTs. The surface functionalization of SWCNTs increases the specific capacitance due to the development of carboxylic groups at the surface as a result of oxidation in nitride acid. An HNO_3-treated sample showed a specific capacitance of 142 F g^{-1} at a current density of 4 A g^{-1}, which is about four times greater than that of an untreated sample (32 A g^{-1}) [77]. Mukhopadhyay *et al* [78] carried out the surface functionalization of SWCNTs by chemical and electrochemical oxidation; the SWCNTs exhibited an improved specific capacitance of 209 F g^{-1} at 50 mA g^{-1} in 1 M H_2SO_4 electrolyte. The enhancement of the specific capacitance was attributed to the contribution of a pseudocapacitive charge storage mechanism resulting from the surface functionalization of the SWCNTs.

Chemical activation of CNTs using KOH is also a well-studied methodology used to improve the specific capacitance. Chemical activation leads to an enormous increase in the SSA [5, 64]. Xie *et al* [79] reported the electrochemical performance of an EDLC based on hierarchical porous carbon microtubes (HPNCTs) produced by a facile carbonization and subsequent KOH activation process using willow catkins. The activation process caused an enhancement of the SSA of about ~1775 m^2 g^{-1} which resulted in a high capacitance of 292 F g^{-1} at 1 A g^{-1} and good capacity retention of 83.5% at 10 A g^{-1}. The same authors made use of the prepared HPNCTs to assemble a coin-type symmetric device with 1 M $LiPF_6$ electrolyte. They measured an energy density of 37.9 Wh kg^{-1} at a power density of 700 W kg^{-1} and 90.6% of capacitance retention over 4000 cycles [79].

It has been reported in the literature that heteroatom doping in CNTs also resulted in a boost to the electrical conductivity and activity of sites on the surface of CNTs for charge storage [80–83]. Heteroatoms, such as those resulting from N doping in CNTs, induce changes in the electronic and crystalline structures of carbons. They also help to improve the surface polarity, electrical conductivity, chemical stability, and ion diffusivity. These changes massively alter the chemical and electrochemical properties of EDLCs. For instance, Xie *et al* [80] prepared N-doped carbon-coated carbon nanotubes (N-C/CNTs) with stable core–shell structures via a one-pot hydrothermal synthesis method. An EDLC fabricated using N-C/CNT electrodes exhibited a specific capacitance of 322.1 F g^{-1} at 1 A g^{-1}

and a specific capacitance of about 120 F g^{-1}, even at a high current density of 20 A g^{-1}. The device also showed good capacitance retention over 10 000 cycles at 5 A g^{-1}. The authors concluded that such high performance was the collective effect of the high electronic conductivity and mechanical strength of the CNTs, the high SSA due to a fine pore structure, and the pseudocapacitive charge storage mechanism introduced by N doping in the carbon layer.

Gueon and Moon [81] synthesized densely packed spherical CNT particles via emulsion-assisted evaporation followed by N doping using melamine. They reported a specific capacitance of 215 F g^{-1} at 0.2 A g^{-1}, which is about three times more than that of undoped CNTs. Furthermore, the samples also showed a capacitance retention of 99% over 1500 cycles at a current density of 1 A g^{-1}. The performance of the N-doped samples was attributed to the well-developed sites with increased activity and good electrical conductivity. Yun *et al* [82] prepared nitrogen and micropores containing carbon nanotubes (NMCNTs) by the carbonization of nitrogen-enriched, polymer-coated carbon nanotubes. The NMCNTs with 9.1 wt% of nitrogen-atom doping showed a specific capacitance of 190.8 F g^{-1} at a scan rate of 5 mV s^{-1}, which is much higher than that of pristine CNTs (48.4 F g^{-1}). The NMCNTs also exhibited a capacitance retention of 94.5% over 10 000 cycles measured at 100 mV s^{-1} in a 1 M H_2SO_4 electrolyte. Moreover, Bulusheva *et al* [83] analyzed the enhanced performance of SCs based on N-doped CNTs using density functional theory calculations for N-CNT models. Their study revealed that nitrogen substitution modifies the electron density distribution in CNTs. The protonation of nitrogen causes a redistribution of electronic charges on neighboring carbon atoms, which further contributes to the enhancement of the specific capacitance. Table 3.3 summarizes the electrochemical performance of EDLCs based on CNTs with different structures.

Thus, the performance of CNT-based supercapacitors is closely related to the CNT type, SSA, porosity, pore size distribution, and post treatment. The porosity and SSA of CNTs can be enhanced either by surface chemical activation or surface functionalization. In addition to this, the ion size and electrode–electrolyte accessibility are also prominent factors that affect the electrochemical performance of CNT-based SCs. The approach of heteroatom doping in CNTs also helps to improve the electrochemical performance. Although CNTs possess high electrical conductivity, their performance is still lower than that of AC-based devices. Novel synthesis methods, structured designs, and hybrid electrodes made from nanocomposites of CNT with graphene, carbon, metal oxides, metal chalcogenides, and conductive polymers may help to improve the specific capacitance and cycle stability of supercapacitors.

In summary, the performance of CNT-based supercapacitors is governed by:

- the type of CNT and its pore size distribution
- the orientation of CNTs and their length
- the electrolyte
- surface functionalization
- chemical activation
- heteroatom doping.

Table 3.3. Electrochemical performance of CNT- and carbon fiber (CF)-based EDLCs.

Working electrode	Electrolyte	Potential window (V)	Specific capacitance (Fg^{-1}) at (@) current density ($A\ g^{-1}$)/ scan rate ($mV\ s^{-1}$)	Power density ($W\ kg^{-1}$)	Energy density ($Wh\ kg^{-1}$)	Retention % (cycles) at (@) current density ($A\ g^{-1}$)/scan rate ($mV\ s^{-1}$)	References
Carbon nanotubes							
SWCNTs	7.5 N KOH	0 to 0.9	180/1 mA cm^{-1}	20 000	~7		[63]
MWCNTs	6 M KOH	0 to 0.8	135/10				[64]
CNT sheet	H_2SO_4	0 to 1	102/10	8000			[68]
Vertically aligned CNT arrays	Ionic liquids	0 to 1	440/5	6300	19.6		[69]
Vertically aligned SWCNTs	$TEABF_4$	1 to 2	75@5	987 000	27	95.7% (2000)	[70]
HNO_3-treated CNTs	0.5 M H_2SO_4	0 to 0.8	142@4				[77]
Functionalized SWCNTs	1 M H_2SO_4	0.2 to 0.7	209@50				[78]
HPC microtubes (SSA~1775.7) /(V_p~ 0.8516)	6 M KOH	0 to 2.8	292@1	7000	37.9	83.5%@10	[79]
Hierarchical porous carbon microtubes (HPNCTs)	1 M $LiPF_6$	0 to 2.8	139@1	700	37.9	90.6% (4000)	[79]
N-doped carbon-coated carbon nanotubes (N-C/CNTs)	2 M KOH	−0.9 to −0.15	322.1@1 and 120@ 20.			214 (10 000)@5	[80]
Nitrogen- and micropore-containing carbon nanotubes (NMCNTs)	1 M H_2SO_4	0 to 1	190.8/5			94.5% (10 000)/100	[82]
Carbon fibers							
CFs without activation	6 M KOH	−1.0 to 0.2	127@50 mA g^{-1}				[84]
	1 M TEABF4/ PC	−1.0 to 0.2	18@50 mA g^{-1}				
	6 M KOH	−1.0 to 0.2	136@50 mA g^{-1}				

(*Continued*)

Table 3.3. (*Continued*)

Working electrode	Electrolyte	Potential window (V)	Specific capacitance (Fg^{-1}) at (@) current density ($A g^{-1}$)/ scan rate ($mV s^{-1}$)	Power density ($W kg^{-1}$)	Energy density ($Wh kg^{-1}$)	Retention % (cycles) at (@) current density ($A g^{-1}$)/scan rate ($mV s^{-1}$)	References
ACFs physically activated by CO_2	1 M $TEABF_4$/ PC	−1.0 to 0.2	63@50 mA g^{-1}				
ACFs chemically activated by KOH	6 M KOH	−1.0 to 0.2	214@50 mA g^{-1}				
	1 M $TEABF_4$/ PC	−1.0 to 0.2	116@50 mA g^{-1}				
CFs from partially hydrolysed cellulose acetate (CA) nanofibers	6 M KOH	−0.9 to 0.1	241.4@1	8410	9.8	99.9% (10 000)	[85]
Nanoporous CFs from (PAN)/ (CA) using $ZnCl_2$	6 M KOH	0 to −0.9	280@0.1 and 223@0.2			96.8% (2000)	[87]
Non-cross-linked CNF	6 M KOH	0 to −0.1	113@1			44.2%@60	[88]
(3D) cross-linked N-doped CNF	6 M KOH	0 to −0.1	163@1			58.9%@ 60	
3D cross-linked N-doped CNF	1 M H_2SO_4	0 to −0.1	214@1				
Non-cross-linked CNF	6 M KOH	0 to −0.1	109@0.2	7400	1.50		
3D cross-linked N-doped CNF	6 M KOH	0 to −0.1	178@0.2	15 000	3.04	97.3% (60 000)	
3D-cross-linked N-doped CNF	1 M H_2SO_4	0 to −0.1	183@1				
N–Ag CNF	1 M H_2SO_4	−0.4 to 0.6	216@1			90% (1000)@2	[89]
CNF/CNTs	6 M KOH	0 to 0.8	149@0.5			90% (1000)	[90]

3.4 Performance of EDLCs based on activated carbon fibers

Another widely studied carbon material for electrodes in EDLCs is carbon fiber (CF). Carbon fibers are carbon materials with a fibrous structure, offering high electrical conductivity with high accessible porosity, higher adsorption rates, and, most importantly, device design flexibility for portable and wearable supercapacitors. They also provide a continuous path for electron conduction and reduce the ion diffusion path, due to their ultrahigh aspect ratio (length to diameter). The structure of CF comprises sheets of carbon atoms similar to those of graphite, which are interlocked instead of being stacked in parallel. Most CFs are manufactured by converting precursors to polymer fibers followed by carbonaceous fibers by heat treatment in the temperature range from 1200 °C to about 3000 °C. Their manufacture involves two main processes: stabilization and carbonization. Stabilization is necessary to prevent the fusion of CFs during carbonization. During stabilization, due to several chemical and physical reactions (e.g., polymerization and dehydrogenation), thermoplastic precursor fibers are converted into highly condensed thermosetting fibers. On the other hand, the carbonization process consists of a heat treatment in an inert gas atmosphere (either nitrogen or argon) to prevent oxidation, which removes impurities.

The most common precursors of CFs are cellulosic rayon, polyacrylonitrile (PAN), and pitch. Depending upon the precursor used to prepare CFs, there are three types, namely, turbostratic CF (T-CF), graphitic CF (G-CF), and hybrid CF (H-CF) (i.e. a composition material made from both G-CF and T-CF). CFs derived from PAN are known as PAN-based carbon fibers and are of the turbostratic type, with wrinkled carbon sheets. They have very good tensile strength compared to other CFs. The CFs developed from pitch are known as pitch-based carbon fibers and are of the graphitic type. They are also referred to as mesophase pitch-based carbon fibers after heat treatment at a high temperature of about 2500 °C. Mesophase pitch-based carbon fibers have high thermal conductivity and a high Young's modulus. All commercially available CFs are PAN-based carbon fibers because PAN can be dissolved in several different solvents. Several methods have been adopted for the preparation of CFs, but chemical vapor deposition and electrospinning are the most widely reported.

Like other carbon materials, to enhance the SSA, ion adsorption capacity, and the pore size, volume, and distribution, CFs can be activated through physical or chemical methods; the resulting fibers are known as activated carbon fibers (ACFs). ACFs are intermediate or inclusive forms of both CFs and ACs. The most common materials used for the synthesis of ACFs are pitch fiber, mesophase pitch, phenolic resins, polyacrylonitrile, and biomass. The selection of raw materials, i.e. the precursor and the activation method, plays an important role in determining the pore size, pore size distribution, surface area, electrical conductivity, and surface functionality, which further affect the electrochemical performance of EDLCs. Several efforts have been made to improve the SSA, electrical conductivity, ion diffusion capacity, pseudocapacitance, etc, of CFs via different approaches.

For instance, Xue *et al* [84] studied the effect of the activation method on the capacitive performance of ACFs derived from phenol–formaldehyde resin fibers. Chemical activation was performed using KOH, while CO_2 was used for physical activation [84]. The results indicated that the SSAs of both activated CFs were greater than those of carbon fibers without activation and had values of 527.0, 939, and 1893 $m^2\ g^{-1}$ for CFs without activation, CFs treated with CO_2, and those treated with KOH, respectively. The resultant EDLC exhibited high specific capacitances of 136 and 214 F g^{-1} at a current density of 50 mA g^{-1} for CFs treated with CO_2 and KOH, respectively, compared to those of CFs without activation, which had a value of about 127 F g^{-1}. The authors also reported electrochemical performance in an organic electrolyte, which yielded capacitances of 18, 63, and 116 F g^{-1} in 1 M $TEABF_4$/PC organic electrolyte for CFs without activation, CFs treated with CO_2, and those treated with KOH, respectively. The maximum capacitance in the case of ACF was attributed to an increased SSA due to the formation and dispersion of pores on its surface, while the enhanced performance of the KOH-treated ACFs compared to the physically activated samples was attributed to inreased surface roughness and a larger surface area.

Cai *et al* synthesized interbonded carbon nanofibers from partially hydrolyzed cellulose acetate nanofibers, which exhibited a specific capacitance of 241.4 F g^{-1} at a current density of 1 A g^{-1}, a maximum power density of ~84.1 kW kg^{-1}, and a very good cyclic stability of about 99.9% over 10 000 cycles. The overall performance was attributed to the interbonded fibrous structure, which offered a short ion diffusion path, good electric conductivity, and well-distributed pores [85].

Lee *et al* prepared pitch-based activated carbon fibers using a steam activation process. The authors reported the impact of activation on the SSA and the total pore volume, which were found to be enhanced from 1520 to 3230 $m^2\ g^{-1}$ and from 0.61 to 1.87 $cm^3\ g^{-1}$, respectively. As an effect of the improved specific surface area and total pore volume, the specific capacitance also increased from 1.1 to 22.5 F g^{-1} at 2 mA cm^{-2} in 1 M $(C_2H_5)_4NBF_4$/propylene carbonate organic electrolyte [86].

Ma *et al* reported the synthesis of nanoporous ultrafine carbon fibers using PAN/cellulose acetate (CA) using $ZnCl_2$ as an activating agent. The authors reported that the mass ratio of CA to PAN significantly influenced the SSA and the pore volume. The carbon fibers prepared with a CA:PAN mass ratio of 1:9, with a surface area of 887 $m^2\ g^{-1}$, and a pore volume of about 0.41 $cm^3\ g^{-1}$ exhibited high specific capacitances of 280 F g^{-1} at 0.1 A g^{-1} and 223 F g^{-1} at 0.2 A g^{-1} in 6 M KOH electrolyte and also exhibited a high capacitance retention of 96.8% over 2000 cycles. The authors concluded that CA had a positive role in surface functionality, helped to widen the pores, and also improved the structural and thermal stability of CFs [87].

Jiang *et al* [88] reported 3D cross-linked nitrogen-enriched porous carbon nanofibers prepared by the electrospinning of a PAN/zinc chloride (PAN/$ZnCl_2$) precursor. A supercapacitor based on PAN/$ZnCl_2$ composite carbon nanofibers (C-CNFs) exhibited a higher specific capacitance than one based on non-cross-linked carbon nanofibers, i.e. pure CNFs (N-CNFs) without the $ZnCl_2$ additive. The surface morphologies of both CNFs and composite CNFs with PAN/$ZnCl_2$ are shown in high-magnification micrographs in figure 3.5. It can clearly be observed

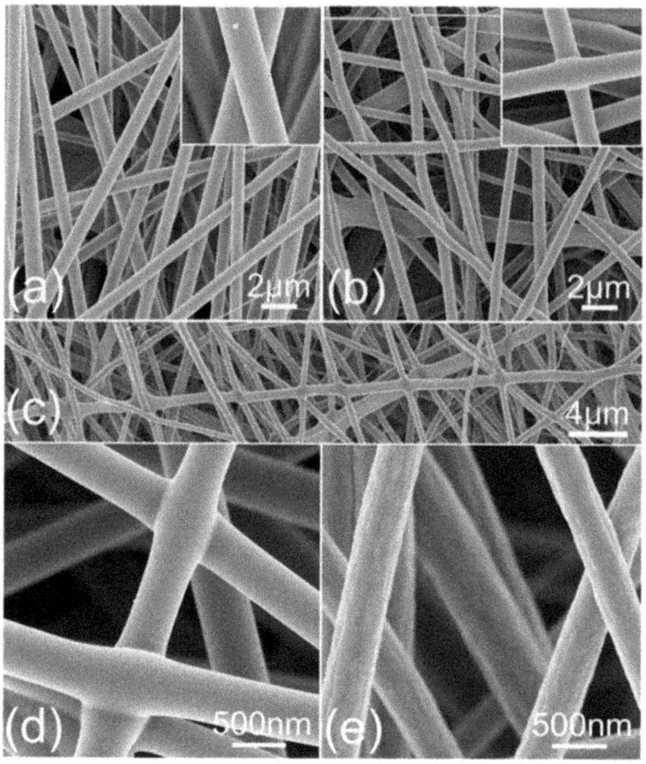

Figure 3.5. SEM micrographs of: (a) C-CNFs and (b) pre-oxidized composite nanofibers. (c) Low-magnification SEM images of C-CNFs. High-magnification SEM images of: (d) C-CNFs and (e) N-CNFs. Reprinted from [88], Copyright (2019), with permission from Elsevier.

from figures 3.5(d) and (e) that the PAN/$ZnCl_2$ composite nanofibers have melted together to form a cross-linked structure because of the cyclization of the nitrile groups in the PAN during the pre-oxidation process at a temperature of about 150 °C, while the pure CNFs are just stacked on each other without any bonding. The SSA of C-CNFs showed a high Brunauer–Emmett–Teller (BET) surface area of 520 m^2 g^{-1}, compared to 107 m^2 g^{-1} for N-CNFs. Similarly, the electrical conductivity of the C-CNFs was about 2.33 S cm^{-1}, which was also more than the value of 1.50 S cm^{-1} measured for N-CNFs. The authors argued that the $ZnCl_2$ additive played a significant role in the pore formation and cross-linking of CFs, which resulted in an enhanced specific surface area and fast charge transfer throughout the cross-linked fibrous network, respectively. The electrochemical performances of CNFs with and without the composite are illustrated in figure 3.6.

Figure 3.6(a) displays the cyclic voltammetry (CV) curves of (a) N-CNFs and (b) C-CNFs in 1 M H_2SO_4 acidic electrolyte at a scan rate 10 mV s^{-1} and for a potential range of −1.0 to 0.0 V. Both electrodes have nearly rectangular curves, which confirms the presence of EDLC charge storage mechanism characteristics. The area of the CV curve is greater for C-CNFs, hence, the C-CNFs delivered more

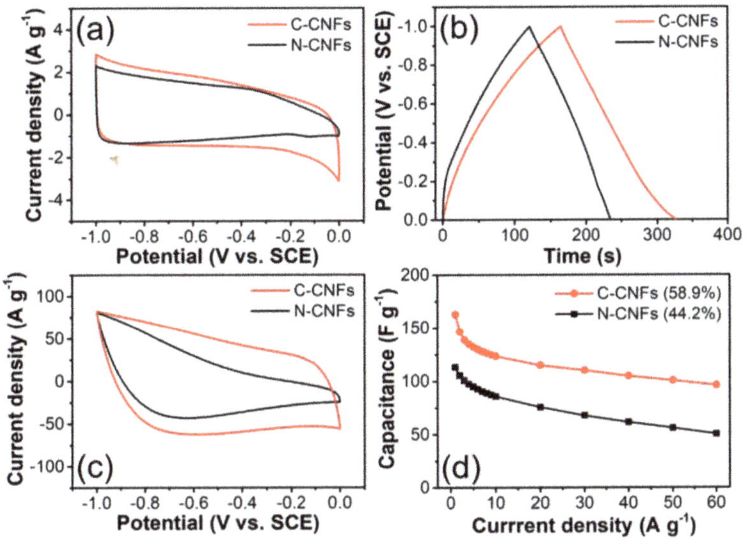

Figure 3.6. Electrochemical performances of C-CNFs and N-CNFs in a three-electrode test: (a) CV curves at a scan rate of 10 mV s^{-1}, (b) GCD plots of C-CNFs and N-CNFs at a current density of 1 A g^{-1}, (c) CV curves at 500 mV s^{-1}, and (d) specific capacitance at current densities in the range of 1–60 A g^{-1}. Reprinted from [88], Copyright (2019), with permission from Elsevier.

capacitance than the N-CNFs. At a high scan rate of 500 mV s^{-1} in figure 3.6(c), no change is observed in the CV curve of the C-CNFs, but the shape of the CV curve for the N-CNFs deviates from the shape obtained at a scan rate 10 mV s^{-1}, which means that the C-CNFs have better rate capability. The specific capacitance and cyclic performance of both samples are estimated using GCD at current densities in the range of 1 to 60 A g^{-1}. Figure 3.6(b) compares the GCD curves for both N-CNFs and C-CNFs in 1 M H_2SO_4 at 1 A g^{-1}. The shape of the GCD curves is close to triangular for both samples, which simply means that both show ideal charging and discharging behavior. The C-NCF electrode shows a longer discharging time than that of the N-CNF electrode because of the high surface area developed by $ZnCl_2$ activation. The C-CNFs exhibit a high specific capacitance of 163 F g^{-1} at a current density of 1 A g^{-1}, compared to that of the N-CNFs (113 F g^{-1}). The variation of specific capacitance (C_s) at different scan rates is shown in figure 3.6(d). It can be observed that C_s decreases with the scan rate and shows retentions of 58.9% and 44.2% at 60 A g^{-1} relative to that at 1 A g^{-1} for C-CNFs and N-CNFs, respectively. The improvement in rate capability was attributed to fast charge transfer through the cross-linked 3D porous structure developed by the $ZnCl_2$ additive. The same study also reported the effects of acidic and alkaline electrolytes on the electrochemical performance of C-CNF-based EDLCs. The specific capacitance of 214 F g^{-1} at 1 A g^{-1}, in 1 M H_2SO_4 acidic electrolyte was found to be better than that in 6 M KOH alkaline electrolyte, which was about 163 F g^{-1} at 1 A g^{-1} . The improved performance in the acidic electrolyte was attributed to the fact that the protons in the acidic electrolyte induced a redox reaction in the functional groups [88].

Recently, Gaminian *et al* studied the capacitive performance of nitrogen-doped silver–carbon nanofibers prepared by the pyrolysis of a regenerated cellulose nanofiber precursor. Interestingly, they used bio-inspired dopamine, which simultaneously served as the precursor, provided reduction and adhesion, and functioned as a binding and stabilizing agent. The nitrogen-doped silver–carbon nanofibers exhibited a high electrical conductivity of 6.41 S cm^{-1}, which was much larger than that of pure CNFs (1.41 S cm^{-1}) and a specific capacitance of 216 F g^{-1} at 1 A g^{-1} in 1 M H_2SO_4 electrolyte with excellent stability over 1000 cycles at 2 A g^{-1} . The observed value of specific capacitance for the nitrogen-doped silver–carbon nanofibers was about four times higher than that of pure CNFs, which was attributed to the higher electrical conductivity of Ag and the pseudocapacitance effect of N doping [89].

Li *et al* prepared a cellulose-derived hybrid carbon nanofiber/carbon nanotube (CNF/CNT) electrode. The SSA of the hybrid CNF/CNTs was found to be 1211 m^2 g^{-1}, which is much higher than the value of 712 m^2 g^{-1} measured for pristine CNFs. As a result of the increase in the surface area, the hybrid CNF/CNTs exhibited a high specific capacitance of 149 F g^{-1} at a current density of 0.5 A g^{-1}, which is about 15% more than that of pristine CNFs. An EDLC with the hybrid CNF/CNT electrode showed a capacitance retention of 90% over 1000 cycles [90]. Table 3.3 summarizes the electrochemical performance of CF-based EDLCs.

Thus, in summary, CFs in different forms, such as activated carbon fibers, carbon nanofibers, porous carbon nanofibers, doped carbon nanofibers, etc, are used extensively as electrodes in EDLCs. Various efforts, in terms of physical and chemical activation, N doping, and optimizing the activation temperature, have been made to enhance the specific surface area, and hence, the electrochemical performance of CNF-based EDLCs. In spite of all these efforts, CFs exhibit a maximum specific capacitance of about 300 F g^{-1}. Composites of CFs with other types of carbon may help to enhance the specific capacitance beyond this limit.

3.5 Performance of EDLCs based on graphene and its composites

Most research in the present era of science and technology is directly or indirectly focused on either graphene or its composite materials, both of which are versatile in applications. Graphene has received enormous attention because of its unique mechanical, elastic, physical, chemical, electrical, thermal, and electrochemical properties. Graphene takes the form of a sheet-like layered structure composed of sp^2 hybridized carbon atoms in a honeycomb structure with alternating C–C and C=C bonds. From an electrochemical charge storage device point of view, it possesses numerous intrinsic properties, such as a large theoretical surface area (2600 m^2 g^{-1}), a flexible structure, high electrical conductivity (2×10^3 S cm^{-1}), good thermal conductivity (5×10^3 W m^{-1} K^{-1}), good mechanical strength (~1 TPa), and high charge carrier mobility (2×10^5 cm^2 V^{-1} s^{-1}) [91–98].

Furthermore, the 2D structure of graphene offers an open pore structure and also decreases the electrode thickness, which leads to a larger potential window [93, 99]. Because of these properties, it has a high estimated theoretical specific capacitance of 550 F g^{-1} [100].

Figure 3.7 shows different forms of graphene used for supercapacitor applications. For EDLCs, graphene electrodes have been reported with different structures, such as nanosheets, foam, hollow spheres, hydrogel, aerogel, and 3D mesoporous structures, intended to enhance its electrochemical performance.

Even though graphene possesses high electrical conductivity, a high surface area, and high theoretical specific capacitance, in practical terms, its performance is far from the predicted values, due to self-restacking. The strong π-π interaction between graphene layers causes a decrease in the SSA and also affects the ion and charge transfer kinetics, which results in poor electrochemical performance [96, 98, 101].

In order to overcome this drawback of pristine graphene, several efforts have been made to modify graphene by oxidization, activation, doping, intercalation of other nanoparticles in its layers, etc. Graphene can be synthesized using several routes, such as the exfoliation method, the templating technique, chemical vapour deposition, arc discharge, molecular beam epitaxy, oxidation–reduction, electrically assisted synthesis, the solvothermal method, pyrolysis, epitaxial growth, and a biomass green synthesis route. Typically, in the oxidation–reduction method, graphite is oxidized to produce graphite oxide using Hummers' method and then graphene oxide (GO) is synthesized by exfoliation or by simple sonication. Furthermore, chemically derived graphene can be obtained by reducing GO, either using a reducing agent (such as hydrazine hydrate) or by a thermal method [102–108]. In the CVD process, a carbon source is used as the precursor to deposit graphene layers onto different metallic or non-metallic substrates. In general, hydrocarbon precursors are used, which are subsequently pyrolyzed to carbon radicals before deposition onto the substrate [108].

Several reports showing the use of graphene in different forms for supercapacitor applications are available in the literature.

Stoller *et al* were the first to report the electrochemical performance of a supercapacitor using graphene; they called their graphene 'chemically modified graphene.' The SSA of a single graphene sheet was measured at about 2630 $m^2\ g^{-1}$, and it exhibited a specific capacitance of 135 F g^{-1} in an aqueous KOH electrolyte [109]. The same group reported a specific capacitance for graphene of about 99 F g^{-1} in an organic electrolyte.

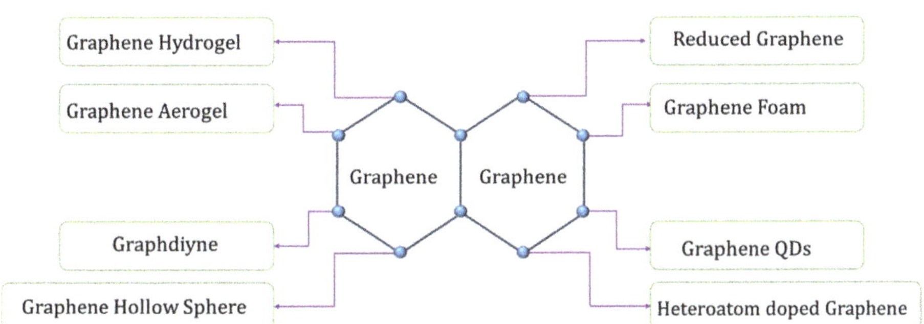

Figure 3.7. Schematic representation of different graphene structures used in supercapacitor applications.

At almost the same time, Vivekchand *et al* prepared graphene by the exfoliation of graphitic oxide, which exhibited a high operating potential window of up to 3.5 V with a specific capacitance and an energy density of about 75 F g^{-1} and 31.9 Wh kg^{-1}, respectively, in an ionic liquid ($PYR_{14}TFSI$). The same study also reported a specific capacitance of 117 F g^{-1} at 100 mV s^{-1} in a 1 M H_2SO_4 electrolyte [110].

As stated earlier, graphene sheets suffer from self-restacking because of van der Waals interactions between adjacent sheets, which not only reduce the SSA, but also limit the charge kinetics and ion diffusion process, resulting in irreversible capacity loss and a reduction in energy density and coulombic efficiency [111]. In the past few decades, graphene oxide has emerged as a promising starting material for the synthesis of functional graphene nanostructures.

For instance, Du *et al* prepared graphene sheets with mesopores from natural graphite via oxidation and rapid heating processes, which exhibited a specific capacitance of 150 F g^{-1} at 0.1 A g^{-1} in a 30% KOH aqueous electrolyte and also maintained stable capacity over 500 charge–discharge cycles [112].

Wang *et al* prepared graphene by the reduction of graphene oxide sheets using gas-based hydrazine as a reducing agent, which exhibited a maximum specific capacitance of 205 F g^{-1} with a power density of 10 kW kg^{-1} and an energy density of 28.5 Wh kg^{-1} in an aqueous electrolyte, with a good retention of ~90% over 1200 cycle [113].

The requirement for high temperatures for thermal exfoliation can be diminished by the use of a microwave treatment in vacuum [112, 114]. Zhu *et al* demonstrated the electrochemical performance of graphitic oxide prepared by microwave-assisted exfoliation and reduction, which exhibited a specific capacitance of 191 F g^{-1} at 0.15 A g^{-1} in a 5 M KOH electrolyte [114].

Several studies have reported the synthesis of different graphene-like nanostructures from biowaste natural resources, such as food, agricultural waste, and wood waste together with activating agents. Recently, Purkait *et al* [115] synthesized layered graphene from peanut shell using KOH as an activating agent. The prepared nanosheets had a high SSA of 2070 m^2 g^{-1} and a pore volume about 1.33 cm^3 g^{-1}; they exhibited a specific capacitance, an energy density and a power density of about 186 F g^{-1} at 0.5 A g^{-1}, 58.25 Wh kg^{-1}, and 37.5 kW kg^{-1}, respectively, in a 1 M H_2SO_4 electrolyte, with a large potential window of 1.5 V (– 0.4 V to 1.1 V) [115].

Moreover, nanostructured graphene oxide has been synthesized with different morphologies such as foam, hollow spheres, and 3D porous structures in order to improve the electrochemical performance of devices [116–118]. Zhao *et al* reported polypyrrole-mediated graphene foam electrodes with a specific capacitance of 151 F g^{-1} at 1.5 A g^{-1} in a 3 M $NaClO_4$ aqueous electrolyte [116].

Furthermore, Xu *et al* prepared a holey 3D framework of graphene sheets, which exhibited a high specific capacitance of 298 F g^{-1} at 1 A g^{-1} in an organic electrolyte with an energy density of 35 Wh kg^{-1} and good capacitance retention of 91% over 10 000 cycles at a high current density of 20 A g^{-1} [117]. The authors argued that the hierarchically porous structured 3D-holey graphene electrode enhanced the ion diffusion process by decreasing the ion transport path, which resulted in an enhancement of the specific capacitance. In another report, Kim *et al* prepared porous graphene-derived carbons with hierarchical pore structures. Mesopores were introduced by chemical activation using KOH, and micropores were introduced by the use of graphene-based hollow spheres. A fabricated EDLC showed a specific capacitance of 174 F g^{-1} at 4.1 A g^{-1} with an energy density of 74 Wh kg^{-1} and a power density of 338 kW kg^{-1} in an ionic liquid electrolyte. The authors reported that such a hierarchical pore structure not only improved the surface area value to 3290 m^2 g^{-1} but also supported the ion diffusion process [118].

Choi *et al* demonstrated high-performance supercapacitors using 3D macroporous embossed chemically modified graphene, which showed a high specific capacitance of 202 F g^{-1} at 1 A g^{-1}. Their electrode also showed a good capacity retention of 98.5% over 1000 cycles at 35 A g^{-1} in a 1 M Na_2SO_4 electrolyte, which was much greater than that of an electrode made from chemically modified graphene (93 F g^{-1} and 87.8%). The enhanced performance was attributed to fast ionic transport because of the large surface area offered by the hierarchical macroporous structure [119].

Another way to improve the electrochemical performance of graphene-based EDLCs is to dope graphene with heteroatoms, such as nitrogen (N), boron (B), sulfur (S), phosphorous (P), etc. Heteroatoms act as electron donors and acceptors, and hence modify the host electronic structure, resulting in changes in the chemical and physical characteristics of graphene. For example, nitrogen has almost same size and electronic configuration as carbon, but it has an extra valence electron, which develops new energy levels in graphene's electronic band structure. Such a modification enhances the surface activity and electrical conductivity of graphene [120–123]. As reported in the literature, an N-doped graphene oxide supercapacitor showed a specific capacitance of 280 F g^{-1}, which is about four times larger than that of pristine graphene [120].

Recently, Liu *et al* prepared N-doped graphene by annealing pristine graphene in ammonia at 500 °C for 1 h under ambient pressure, which showed specific capacitances of 390 and 354 F g^{-1} at 5 and 10 A g^{-1} in 6 M KOH, respectively. The sample also exhibited a good retention of 95% over 5000 cycles at a high current density of 50 A g^{-1} between 0 and −0.8 V. The authors stated that nitrogen doping of the graphite lattice increased the density of states near the Fermi level [124].

Zhao *et al* synthesized an N-doped 3D graphene framework by the reduction of graphene oxide using a nitrogen precursor by the hydrothermal method. The resultant sample revealed specific capacitances of 484 and 415 F g^{-1} at 1 and 100 A g^{-1} in a 1 M $LiClO_4$ electrolyte and also maintained a stable capacity over 1000 cycles at 100 A g^{-1} [125]. Like nitrogen, boron also has the same size as carbon, and hence also improved the performance of an EDLC which showed a specific capacitance of 200 F g^{-1} at 0.1 A g^{-1} in a 6 M KOH electrolyte [126].

In past few years, graphene-based self-assembled 3D porous nanostructures, such as hydrogel and aerogel have emerged as promising electrode materials for electrochemical charge storage devices. To date, a large number of reports have been published on the subject of graphene aerogel and its modified forms, in terms of doping and activation.

For instance, Xu *et al* prepared a self-assembled graphene hydrogel with good cycling stability and remarkable mechanical flexibility via a convenient one-step hydrothermal method. The electrode made using this graphene showed a high electrical conductivity of about 5×10^3 S cm^{-1} and a specific capacitance of 160 F g^{-1} at 1 A g^{-1} in a 5 M KOH electrolyte. In the same report, the authors argued that the cross-linked graphene hydrogel was much stronger than conventional self-assembled hydrogels, and hence had a high thermal stability [127].

Jung *et al* reported porous graphene aerogel with a specific capacitance of 325 F g^{-1} at 1 A g^{-1} and an energy density of 45 Wh kg^{-1} in a 0.5 M H_2SO_4 aqueous electrolyte; the aerogel had a capacitance retention of about ~98% over 5000 cycles. According to the authors, the temperature used in the freeze-drying step of the aerogel formation affected the pore size and distribution [128].

Zhang *et al* studied the capacitive performance of a 3D superelastic graphene aerogel nanosheet and obtained an electrical conductivity as high as ~1000 S m^{-1}, with specific capacitances of 245 and 201 F g^{-1} at 1 and 20 A g^{-1} respectively, and a retention of 92% over 10 000 cycles in a 1 M H_2SO_4 electrolyte [129].

Ye *et al* fabricated a supercapacitor with a binder-free electrode by the direct deposition of 3D graphene aerogel on a nickel foam, which exhibited a specific capacitance of 366 F at 2 A g^{-1} in a 6 M KOH electrolyte and a capacitance retention of about 60% at a high current density of 10 A g^{-1} over 2000 cycles at a potential window between 0 and −1.0 V. Its improved performance was attributed to the high mass, hierarchical porosity, and high conductivity of the electrode. The authors stated that the direct growth of the 3D graphene aerogel network structure eased the ion diffusion process and also improved the electrical conductivity, which resulted in enhanced performance [130].

The doping of graphene aerogels also has an effect on their electrochemical properties, which is similar to the case of graphene nanosheets. Xu *et al* reported the enhanced electrochemical performance of a supercapacitor based on 3D N-doped holey graphene aerogel prepared by the hydrothermal method.

The self-assembled 3D interconnected holey graphene aerogel (HGA) was obtained using H_2O_2, which created carbon vacancies from the oxygenated carbon species. Furthermore, nitrogen-doped holey graphene aerogel (NHGA) was prepared using hydrazine hydrate as the nitrogen source. Figure 3.8 shows a schematic of the preparation and structure of the HGA and NHGA samples [131]. The study reported a high SSA of 446 m^2 g^{-1} and a low charge transfer resistance of 0.4 Ω for 3D N-doped holey graphene aerogel. Figure 3.9 displays the electrochemical performances of the HGA and NHGA electrodes. From figures 3.9(a) and (b), it can be observed that the CV curve of NHGA has small superimposed pairs of wide Faraday peaks and is more symmetrical and rectangular than that of HGA. According to the report, the peaks in the CV curve of NHGA are significant signs of additional pseudocapacitance due to N doping and the presence of residual surface oxygen-containing functional groups. Figures 3.9(c) and (d) show the GCD curves of NHGA and HGA, respectively. The specific capacitances of NHGA at different current densities can be seen to be larger than those of HGA. The resulting novel 3D network exhibits an excellent specific capacitance of 318.3 F g^{-1} at 0.5 A g^{-1} in a 6 M

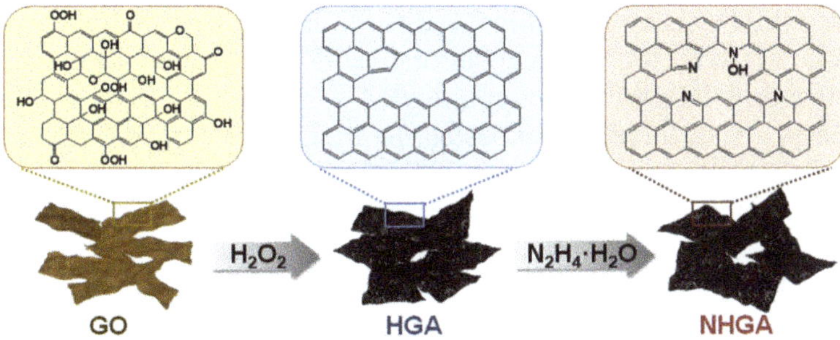

Figure 3.8. Schematic illustration of the preparation and structure of HGA and NHGA samples. Reprinted from [131], Copyright (2019), with permission from Elsevier.

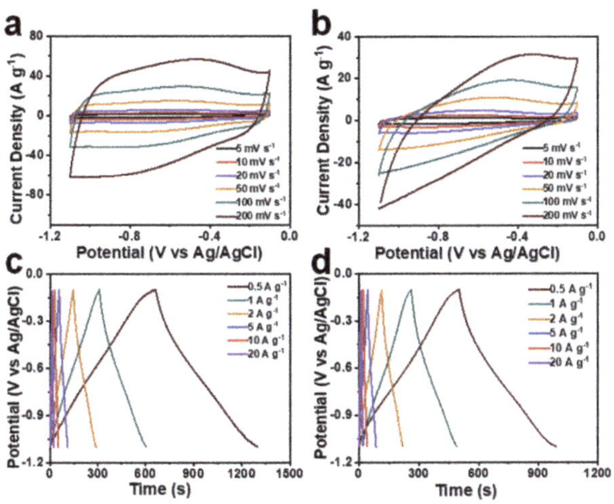

Figure 3.9. CV curves of: (a) NHGA and (b) HGA; GCD plots of: (c) NHGA and (d) HGA at different current densities. Reprinted from [131], Copyright (2019), with permission from Elsevier.

KOH electrolyte with an energy density of 60.3 Wh kg^{-1} and a good capacity retention of 98.4% at 2 A g^{-1} over 10 000 cycles [131].

Li *et al* fabricated boron-doped graphene aerogels by a simple hydrothermal method; the aerogels showed a specific capacitance of 308.3 F g^{-1} at 1 A g^{-1} and a good capacitance retention of about 92% after 5000 cycles at 1 A g^{-1}, which was higher than that of undoped graphene aerogels (86%) [132]. Likewise, some trials of co-doped graphene aerogels have also been reported. However, nitrogen and sulfur co-doping resulted in better performance than N/P doping [133, 134].

Enhanced supercapacitance performance, compared to that of graphene, has also been reported for graphene quantum dots (GQDs). Hassan *et al* synthesized edge-enriched activated GQDs using KOH as an activating agent; they exhibited a specific capacitance of 172 F g^{-1}, which was much higher than that of layered graphene sheets (63 F g^{-1}). The enhanced performance was attributed to decreased charge transfer resistance and a fast ion diffusion process in the activated GQDs [135].

Graphdiyne (GDY) is another advanced form of graphene that has emerged as a novel electrode material for supercapacitors. GDY is a combination of sp and sp^2 hybrid carbon, and its enriched π-conjugated systems may help the fast charge transfer process; it is therefore expected to yield an improved electrochemical performance [136–139]. The first report of GDY by Krishnamoorthy *et al* described a specific capacitance of 71.4 F g^{-1} at a current density of 3.5 A g^{-1} in a Na_2SO_4 electrolyte [137]. Shang *et al* developed N-doped GDY with a microporous structure, which displayed a high specific capacitance of 247 F g^{-1} at 0.2 A g^{-1}, an energy density of 8.55 Wh kg^{-1}, and a power density of 18.7 kW kg^{-1} in a 7 M KOH aqueous electrolyte. The sample also exhibited a high retention of 95.6% at 5 A g^{-1} over 3000 cycles [140]. Table 3.4 summarizes the electrochemical performance of graphene-based EDLCs.

In summary, the performance of graphene-based supercapacitors is governed by:

- the stacking of the graphene layers
- the electrolyte
- surface activation
- 3D porous structures
- heteroatom doping.

Thus, in conclusion, compared with AC, CNTs, and CFs, graphene is the most versatile material; the performance of graphene-based supercapacitors is mainly governed by the layer stacking process, the electrical conductivity, and the hierarchy of the porous network structure. Several efforts have been made to improve the porosity, conductivity, and SSA of graphene, namely, surface activation, doping, and 3D porous structures. N doping and 3D porous structures have been found to be more beneficial for improved electrochemical performance. Composites of graphene in different forms may be used as novel electrodes in future commercialized supercapacitors.

3.6 Summary

The objectives of this chapter are to explore various key factors that affect the performance of EDLCs based on carbon materials, to explain the role of each key factor, and to describe the optimizations made to enhance their performance. The relevant facts are presented here from related studies carried out in the last two decades. The selection of electrode materials, their combinations, and their synthesis or fabrication techniques play a vital role in enhancing the electrochemical performance of EDLCs.

In general, an SC electrode material must have following characteristics:

- thermal stability
- structural stability
- chemical and electrochemical stability
- high electrical conductivity
- suitable surface wettability
- high SSA

Table 3.4. Electrochemical performance of EDLCs based on graphene and its different structures.

Working electrode (SSA ($m^2\ g^{-1}$))/(pore volume, V_p ($cm^3\ g^{-1}$))	Electrolyte	Potential window (V)	Specific capacitance (Fg^{-1}) at (@) current density ($A\ g^{-1}$)/scan rate ($mV\ s^{-1}$)	Power density ($W\ kg^{-1}$)	Energy density ($Wh\ kg^{-1}$)	Retention % (cycles) at (@) current density ($A\ g^{-1}$)/scan rate ($mV\ s^{-1}$)	Ref.
Chemically modified graphene (SSA~2630)	5.5 M KOH 1 M $TEABF_4$	0 to 1 0 to 2.5	135@10 m Ag^{-1} and 101/20 99 @10 mA g^{-1}			97 Fg^{-1}/400	[109]
Graphene made by the exfoliation of GO	Ionic $PYR_{14}TFSI$ 1 M H_2SO_4	0 to 3.5 0 to 1	75/100 117/100		31.9		[110]
Graphene nanosheets (GNSs)	30 wt.% KOH	0.9 to 0.01	150@0.1			100% (500)	[112]
GO using a hydrazine reducing agent	30 wt% KOH	0 to 1	205 @100 m Ag^{-1}	10 000	28.5	90% (1200)	[113]
Microwave-assisted exfoliation and reduction of graphite oxide	5 M KOH	0 to 0.9	191@0.15			9%@0.6	[114]
Graphene from waste biomass (SSA~2070) /(V_p ~1.33)	1 M H_2SO_4	−0.4 to −1.1	186@0.5	37 500	58.25		[115]
Polypyrrole-mediated graphene foam	3 M $NaClO_4$	0 to 2.5	151@1.5			50% (1000)	[116]
Hollow graphene	1-ethyl-3-methylimidazolium tetrafluoroborate/ acetonitrile ($EMIMBF_4$/AN)	0 to 3.5	298@1		35	91% (10 000) @20	[117]

Graphene spheres	Ionic electrolytes such as [EMMI][TFSI] and [BMIM][BF_4] in acetonitrile	0 to 3.5	174@4.1	338 000	74	94% (1000)	[118]
3D macroporous embossed graphene	1 M Na_2SO_4	0 to 1	202@1			98.5% (1000)@35	[119]
N-doped graphene	6 M KOH	0 to 0.8	390@5 and 354@10			95% (5000)@50	[124]
3D N-doped graphene	1 M $LiClO_4$	0 to 0.8	484@1 and 415@100			100% (1000)@100	[125]
B-doped rGO	6 M KOH	−1 to 0	200@0.1	~10 000	~5.5	95% (4500)	[126]
Graphene hydrogel	5 M KOH	0 to 1	160@1				[127]
Graphene aerogel	0.5 M H_2SO_4	0 to 1	325@1	7	45	~98% (5000)	[128]
3D superelastic graphene aerogel nanosheet	1 M H_2SO_4	0 to 1	245@1			92% (10 000)@15	[129]
3D graphene aerogel on Ni foam	6 M KOH	−1 to 0	366@2			60% (2000) @10	[130]
3D NHGA	6 M KOH	−1.1 to 0.1	318.3 @ 0.5		60.3	98.4% (10 000)@2	[131]
Boron-doped graphene aerogel	6 M KOH	−1 to 0	308.3 @1			92% (5000) @1	[132]
N–S co-doped graphene	$EMIMBF_4$	0 to 4	182.3 @1		101		[133]
N/P doped rGO arogel	6 M KOH	−0.1 to −0.9	165@0.1			80% (2000)@0.5	[134]
Graphene quantum dots	1 M H_2SO_4	−0.2 to 0.7	236@1			98.8% @10	[135]
GDY	Na_2SO_4	−0.8 to 0	71.4@3.5			97% (1000)	[137]
GDY with a microporous structure	7 M KOH	0 to 1	247@0.2	18 700	8.55	95.6% (3000)@5	[140]

- corrosion resistance
- low cost and environment friendly
- able to transfer the faradaic charge.

It has been consistently reported that carbon electrodes with mesopores exhibit higher specific capacitance due to increased activity at the electrochemical sites. Pore sizes that are too small lead to higher equivalent resistance and less access to the electrolyte. Conversely, effective distributions of the micro/mesopores of the electrode material yield good retention capability with high power density, due to the increased area that can be accessed by the electrolyte, which further provides an easy pathway for charge transportation kinetics at the interfaces. Several features of carbon and its derivatives, such as their good mechanical strength, high surface area, low cost, abundant resources, high corrosion resistance, acceptable electronic conductivity, and ease of scaling up are important for their use in supercapacitors.

The electrochemical performance of SC electrodes based on carbon and its derivatives has been found to be improved by:

- suitable combinations of pore size and pore size distribution
- optimization of the pore size according to the electrolyte used
- the use of non-aqueous electrolytes to obtain a better potential window
- surface functionalization
- optimization of the temperature and atmosphere used for the carbonization process
- physical and chemical activation control parameters
- N doping
- the precursor-to-activating-agent mass ratio
- layer stacking in the case of graphene
- orientation in the case of CNTs.

The properties and electrochemical performance analysis of different carbon materials and their nanocomposites allow us to understand the extent to which these materials are suitable for energy storage applications.

References

[1] Frackowiak E and Beguin F 2001 Carbon materials for the electrochemical storage of energy in capacitors *Carbon* **39** 937–50

[2] Kim B K, Sy S, Yu A and Zhang J 2015 Electrochemical supercapacitors for energy storage and conversion *Handbook of Clean Energy Systems* (New York: Wiley) pp 1–25

[3] Yu Z, Tetard L, Zhai L and Thomas J 2015 Supercapacitor electrode materials: nanostructures from 0 to 3 dimensions *Energy Environ. Sci.* **8** 702–30

[4] Iro Z S, Subramani C and Dash S S 2016 A brief review on electrode materials for supercapacitor *Int. J. Electrochem. Sci.* **11** 10628–43

[5] Pandolfo A G and Hollenkamp A F 2006 Carbon properties and their role in supercapacitors *J. Power Sources* **157** 11–27

[6] Liu C, Li F, Ma L P and Cheng H M 2010 Advanced materials for energy storage *Adv. Mater.* **22** E28-62

[7] Chen Y, Zhang X, Zhang H, Sun X, Zhang D and Ma Y 2012 High-performance supercapacitors based on a graphene–activated carbon composite prepared by chemical activation *RSC Adv.* **2** 7747–53
[8] Yang Z, Ren J, Zhang Z, Chen X, Guan G, Qiu L, Zhang Y and Peng H 2015 Recent advancement of nanostructured carbon for energy applications *Chem. Rev.* **115** 5159–223
[9] Moreno-Fernandez G, Ibañez J, Rojo J M and Kunowsky M 2017 Activated carbon fiber monoliths as supercapacitor electrodes *Adv. Mater. Sci. Eng.* **2017**
[10] Peng H, Yao B, Wei X, Liu T, Kou T, Xiao P, Zhang Y and Li Y 2019 Pore and heteroatom engineered carbon foams for supercapacitors *Adv. Energy Mater.* **9** 1803665
[11] Yang X, Fei B, Ma J, Liu X, Yang S, Tian G and Jiang Z 2018 Porous nanoplatelets wrapped carbon aerogels by pyrolysis of regenerated bamboo cellulose aerogels as supercapacitor electrodes *Carbohydrate Polym.* **180** 385–92
[12] Raymundo-Pinero E, Kierzek K, Machnikowski J and Béguin F 2006 Relationship between the nanoporous texture of activated carbons and their capacitance properties in different electrolytes *Carbon* **44** 2498–507
[13] Barbieri O, Hahn M, Herzog A and Kötz R 2005 Capacitance limits of high surface area activated carbons for double layer capacitors *Carbon* **43** 1303–10
[14] Kleszyk P, Ratajczak P, Skowron P, Jagiello J, Abbas Q, Frąckowiak E and Béguin F 2015 Carbons with narrow pore size distribution prepared by simultaneous carbonization and self-activation of tobacco stems and their application to supercapacitors *Carbon* **81** 148–57
[15] Andas J and Satar N A A 2018 Synthesis and characterization of tamarind seed activated carbon using different types of activating agents: a comparison study *Mater. Today Proc.* **5** 17611–17
[16] Li S, Han K, Li J, Li M and Lu C 2017 Preparation and characterization of super activated carbon produced from gulfweed by KOH activation *Microporous Mesoporous Mater.* **243** 291–300
[17] Yu S, Sun N, Hu L, Wang L, Zhu Q, Guan Y and Xu B 2018 Self-template and self-activation synthesis of nitrogen-doped hierarchical porous carbon for supercapacitors *J. Power Sources* **405** 132–41
[18] Moussavi G, Alahabadi A, Yaghmaeian K and Eskandari M 2013 Preparation, characterization and adsorption potential of the NH_4Cl-induced activated carbon for the removal of amoxicillin antibiotic from water *Chem. Eng. J.* **217** 119–28
[19] Wei Q, Chen Z, Cheng Y, Wang X, Yang X and Wang Z 2019 Preparation and electrochemical performance of orange peel based-activated carbons activated by different activators *Colloids Surf.,* A **574** 221–7
[20] Oliveira L C, Pereira E, Guimaraes I R, Vallone A, Pereira M, Mesquita J P and Sapag K 2009 Preparation of activated carbons from coffee husks utilizing $FeCl_3$ and $ZnCl_2$ as activating agents *J. Hazard. Mater.* **165** 87–94
[21] Legrouri K, Khouya E, Ezzine M, Hannache H, Denoyel R, Pallier R and Naslain R 2005 Production of activated carbon from a new precursor molasses by activation with sulphuric acid *J. Hazard. Mater.* **118** 259–63
[22] Gao Y, Yue Q, Xu S and Gao B 2015 Activated carbons with well-developed mesoporosity prepared by activation with different alkali salts *Mater. Lett.* **146** 34–6
[23] Qiu D, Guo N, Gao A, Zheng L, Xu W, Li M, Wang F and Yang R 2019 Preparation of oxygen-enriched hierarchically porous carbon by $KMnO_4$ one-pot oxidation and activation: mechanism and capacitive energy storage *Electrochim. Acta* **294** 398–405

[24] Zhou S, Chen K, Quan H, Su M, Zeng Z, Pan M, Chen D and Guo L 2019 Molten-$NaNH_2$ activated carbon cloth with high areal capacitance and exceptional rate stability for flexible asymmetric supercapacitors *J. Mater. Sci.* **54** 9111–23

[25] Qu D and Shi H 1998 Studies of activated carbons used in double-layer capacitors *J. Power Sources* **74** 99–107

[26] Endo M, Maeda T, Takeda T, Kim Y J, Koshiba K, Hara H and Dresselhaus M S 2001 Capacitance and pore-size distribution in aqueous and nonaqueous electrolytes using various activated carbon electrodes *J. Electrochem. Soc.* **148** A910

[27] Kierzek K, Frackowiak E, Lota G, Gryglewicz G and Machnikowski J 2004 Electrochemical capacitors based on highly porous carbons prepared by KOH activation *Electrochim. Acta* **49** 515–23

[28] Salitra G, Soffer A, Eliad L, Cohen Y and Aurbach D 2000 Carbon electrodes for double-layer capacitors I. Relations between ion and pore dimensions *J. Electrochem. Soc.* **147** 2486

[29] Raymundo-Piñero E, Leroux F and Béguin F 2006 A high-performance carbon for supercapacitors obtained by carbonization of a seaweed biopolymer *Adv. Mater.* **18** 1877–82

[30] Eliad L, Salitra G, Soffer A and Aurbach D 2001 Ion sieving effects in the electrical double layer of porous carbon electrodes: estimating effective ion size in electrolytic solutions *J. Phys. Chem.* **105** 6880–87

[31] Yamada H, Nakamura H, Nakahara F, Moriguchi I and Kudo T 2007 Electrochemical study of high electrochemical double layer capacitance of ordered porous carbons with both meso/macropores and micropores *J. Phys. Chem.* **111** 227–33

[32] Moriguchi I, Nakahara F, Furukawa H, Yamada H and Kudo T 2004 Colloidal crystal-templated porous carbon as a high performance electrical double-layer capacitor material *Electrochem. Solid State Lett.* **7** A221

[33] Sevilla M and Mokaya R 2014 Energy storage applications of activated carbons: supercapacitors and hydrogen storage *Energy Environ. Sci.* **7** 1250–80

[34] Xia K, Gao Q, Jiang J and Hu J 2008 Hierarchical porous carbons with controlled micropores and mesopores for supercapacitor electrode materials *Carbon* **46** 1718–26

[35] Li L, Wang X, Wang S, Cao Z, Wu Z, Wang H, Gao Y and Liu J 2016 Activated carbon prepared from lignite as supercapacitor electrode materials *Electroanalysis* **28** 243–48

[36] Redondo E, Carretero-González J, Goikolea E, Ségalini J and Mysyk R 2015 Effect of pore texture on performance of activated carbon supercapacitor electrodes derived from olive pits *Electrochim. Acta* **160** 178–84

[37] Luo H, Yang Y, Zhao X, Zhang J and Chen Y 2015 3D sponge-like nanoporous carbons via a facile synthesis for high-performance supercapacitors: direct carbonization of tartrate salt *Electrochim. Acta* **169** 13–21

[38] Boujibar O, Ghamouss F, Ghosh A, Achak O and Chafik T 2019 Activated carbon with exceptionally high surface area and tailored nanoporosity obtained from natural anthracite and its use in supercapacitors *J. Power Sources* **436** 226882

[39] Yin Y, Gao Y, Zhang Y, Li A and Ji G 2019 Synthesis of honeycomb-like hierarchical porous carbon via molten salt pyrolysis in a novel sequencing integration system for high-performance supercapacitors *Microporous Mesoporous Mater.* **278** 195–205

[40] Wang C, Wu D, Wang H, Gao Z, Xu F and Jiang K 2017 Nitrogen-doped two-dimensional porous carbon sheets derived from clover biomass for high performance supercapacitors *J. Power Sources* **363** 375–83

[41] Li Q, Wu X, Zhao Y, Miao Z, Xing L, Zhou J, Zhao J and Zhuo S 2018 Nitrogen doped hierarchical porous carbon through one step activation of bean curd for high-performance supercapacitor electrode *ChemElectroChem.* **5** 1606–14
[42] Li J, Tian L, Liang F, Wang J, Han L, Zhang J, Ge S, Dong L, Zhang H and Zhang S 2019a Molten salt synthesis of hierarchical porous N-doped carbon sub-microspheres for multifunctional applications: high performance supercapacitor, dye removal and CO2 capture *Carbon* **141** 739–47
[43] Jia S, Wei J, Meng X and Shao Z 2020 Facile and friendly preparation of N/S co-doped graphene-like carbon nanosheets with hierarchical pore by molten salt for all-solid state supercapacitor *Electrochim. Acta* **331** 135338
[44] Wu X, Ding B, Zhang C, Li B and Fan Z 2019 Self-activation of nitrogen and sulfur dual doping hierarchical porous carbons for asymmetric supercapacitors with high energy densities *Carbon* **153** 225–33
[45] Su X L, Li S H, Jiang S, Peng Z K, Guan X X and Zheng X C 2018 Superior capacitive behavior of porous activated carbon tubes derived from biomass waste-cotonier strobili fibers *Adv. Powder Technol.* **29** 2097–107
[46] Nam H, Choi W, Genuino D A and Capareda S C 2018 Development of rice straw activated carbon and its utilizations *J. Environ. Chem. Eng.* **6** 5221–29
[47] Song M, Zhou Y, Ren X, Wan J, Du Y, Wu G and Ma F 2019 Biowaste-based porous carbon for supercapacitor: the influence of preparation processes on structure and performance *J. Colloid Interface Sci.* **535** 276–86
[48] Tian Q, Wang X, Xu X, Zhang M, Wang L, Zhao X, An Z, Yao H and Gao J 2018 A novel porous carbon material made from wild rice stem and its application in supercapacitors *Mater. Chem. Phys.* **213** 267–76
[49] Rawal S, Joshi B and Kumar Y 2018 Synthesis and characterization of activated carbon from the biomass of Saccharum bengalense for electrochemical supercapacitors *J. Energy Storage* **20** 418–26
[50] Fu M, Chen W, Zhu X, Yang B and Liu Q 2019 Crab shell derived multi-hierarchical carbon materials as a typical recycling of waste for high performance supercapacitors *Carbon* **141** 748–57
[51] Li Y T, Pi Y T, Lu L M, Xu S H and Ren T Z 2015 Hierarchical porous active carbon from fallen leaves by synergy of K_2CO_3 and their supercapacitor performance *J. Power Sources* **299** 519–28
[52] Yin L, Chen Y, Zhao X, Hou B and Cao B 2016 3-Dimensional hierarchical porous activated carbon derived from coconut fibers with high-rate performance for symmetric supercapacitors *Mater. Des.* **111** 44–50
[53] Misnon I I, Zain N K M, Abd Aziz R, Vidyadharan B and Jose R 2015 Electrochemical properties of carbon from oil palm kernel shell for high performance supercapacitors *Electrochim. Acta* **174** 78–86
[54] Wang C, Wu D, Wang H, Gao Z, Xu F and Jiang K 2018 A green and scalable route to yield porous carbon sheets from biomass for supercapacitors with high capacity *J. Mater. Chem.* **6** 1244–54
[55] Zhang W, Lin N, Liu D, Xu J, Sha J, Yin J, Tan X, Yang H, Lu H and Lin H 2017 Direct carbonization of rice husk to prepare porous carbon for supercapacitor applications *Energy* **128** 618–25
[56] Karazhanov S Z and Raghu S 2018 Synthesis of nanoporous carbon with new activating agent for high-performance supercapacitor *Mater. Lett.* **218** 181–4

[57] Phiri J, Dou J, Vuorinen T, Gane P A C and Maloney T C 2019 Highly porous willow wood-derived activated carbon for high-performance supercapacitor electrodes *ACS Omega* **4** 18108–17
[58] Khan A, Senthil R A, Pan J, Osman S, Sun Y and Shu X 2020 A new biomass derived rodlike porous carbon from tea-waste as inexpensive and sustainable energy material for advanced supercapacitor application *Electrochim. Acta* **335** 135588
[59] Yang L, Qiu J, Wang Y, Guo S, Feng Y, Dong D and Yao J 2020 Molten salt synthesis of hierarchical porous carbon from wood sawdust for supercapacitors *J. Electroanal. Chem.* **856** 113673
[60] He J, Zhang D, Han M, Liu X, Wang Y, Li Y, Zhang X, Wang K, Feng H and Wang Y 2019 One-step large-scale fabrication of nitrogen doped microporous carbon by self-activation of biomass for supercapacitors application *J. Storage Mater.* **21** 94–104
[61] Wang G, Liang R, Liu L and Zhong B 2014 Improving the specific capacitance of carbon nanotubes-based supercapacitors by combining introducing functional groups on carbon nanotubes with using redox-active electrolyte *Electrochim. Acta* **115** 183–8
[62] Bai X, Hu X, Zhou S, Yan J, Sun C, Chen P and Li L 2013 *In situ* polymerization and characterization of grafted poly (3, 4-ethylenedioxythiophene)/multiwalled carbon nanotubes composite with high electrochemical performances *Electrochim. Acta* **87** 394–400
[63] An K H, Kim W S, Park Y S, Moon J M, Bae D J, Lim S C, Lee Y S and Lee Y H 2001 Electrochemical properties of high power supercapacitors using single walled carbon nanotube electrodes *Adv. Funct. Mater.* **11** 387–92
[64] Frackowiak E, Metenier K, Bertagna V and Beguin F 2000 Supercapacitor electrodes from multiwalled carbon nanotubes *Appl. Phys. Lett.* **77** 2421–23
[65] Odom T W, Huang J L, Kim P and Lieber C M 1998 Atomic structure and electronic properties of single-walled carbon nanotubes *Nature* **391** 62–4
[66] Li X, Rong J and Wei B 2010 Electrochemical behavior of single-walled carbon nanotube supercapacitors under compressive stress *ACS Nano* **4** 6039–49
[67] Dillon A C 2010 Carbon nanotubes for photoconversion and electrical energy storage *Chem. Rev.* **110** 6856–72
[68] Niu C, Sichel E K, Hoch R, Moy D and Tennent H 1997 High power electrochemical capacitors based on carbon nanotube electrodes *Appl. Phys. Lett.* **70** 1480–2
[69] Lu W, Qu L, Henry K and Dai L 2009 High performance electrochemical capacitors from aligned carbon nanotube electrodes and ionic liquid electrolytes *J. Power Sources* **189** 1270–7
[70] Kim B, Chung H and Kim W 2012 High-performance supercapacitors based on vertically aligned carbon nanotubes and non-aqueous electrolytes *Nanotechnology* **23** 155401
[71] Shah R, Zhang X and Talapatra S 2009 Electrochemical double layer capacitor electrodes using aligned carbon nanotubes grown directly on metals *Nanotechnology* **20** 395202
[72] Kim Y T, Ito Y, Tadai K, Mitani T, Kim U S, Kim H S and Cho B W 2005 Drastic change of electric double layer capacitance by surface functionalization of carbon nanotubes *Appl. Phys. Lett.* **87** 234106
[73] Lee J Y, An K H, Heo J K and Lee Y H 2003 Fabrication of supercapacitor electrodes using fluorinated single-walled carbon nanotubes *J. Phys. Chem.* **107** 8812–15
[74] Pan H, Li J and Feng Y 2010 Carbon nanotubes for supercapacitor *Nanoscale Res. Lett.* **5** 654–68
[75] Frackowiak E, Delpeux S, Jurewicz K, Szostak K, Cazorla-Amoros D and Beguin F 2002 Enhanced capacitance of carbon nanotubes through chemical activation *Chem. Phys. Lett.* **361** 35–41

[76] Liu C G, Fang H T, Li F, Liu M and Cheng H M 2006 Single-walled carbon nanotubes modified by electrochemical treatment for application in electrochemical capacitors *J. Power Sources* **160** 758–61
[77] Shen J, Liu A, Tu Y, Foo G, Yeo C, Chan-Park M B, Jiang R and Chen Y 2011 How carboxylic groups improve the performance of single-walled carbon nanotube electrochemical capacitors? *Energy Environ. Sci.* **4** 4220–29
[78] Mukhopadhyay I, Suzuki Y, Kawashita T, Yoshida Y and Kawasaki S 2010 Studies on surface functionalized single wall carbon nanotube for electrochemical double layer capacitor application *J. Nanosci. Nanotechnol.* **10** 4089–94
[79] Xie L *et al* 2016 Hierarchical porous carbon microtubes derived from willow catkins for supercapacitor applications *J. Mater. Chem.* **4** 1637–46
[80] Xie C, Yang S, Xu X, Shi J W and Niu C 2020 core–shell structured carbon nanotubes/N-doped carbon layer nanocomposites for supercapacitor electrodes *J. Nanopart. Res.* **22** 1–7
[81] Gueon D and Moon J H 2015 Nitrogen-doped carbon nanotube spherical particles for supercapacitor applications: emulsion-assisted compact packing and capacitance enhancement *ACS Appl. Mater. Interfaces* **7** 20083–9
[82] Yun Y S, Park H H and Jin H J 2012 Pseudocapacitive effects of N-doped carbon nanotube electrodes in supercapacitors *Materials* **5** 1258–66
[83] Bulusheva L G, Fedorovskaya E O, Kurenya A G and Okotrub A V 2013 Supercapacitor performance of nitrogen-doped carbon nanotube arrays *Phys. Status Solidi* B **250** 2586–91
[84] Xue R, Yan J, Liu X, Tian Y and Yi B 2011 Effect of activation on the carbon fibers from phenol–formaldehyde resins for electrochemical supercapacitors *J. Appl. Electrochem.* **41** 1357–66
[85] Cai J, Niu H, Wang H, Shao H, Fang J, He J, Xiong H, Ma C and Lin T 2016 High-performance supercapacitor electrode from cellulose-derived, inter-bonded carbon nanofibers *J. Power Sources* **324** 302–8
[86] Lee H M, Kwac L K, An K H, Park S J and Kim B J 2016 Electrochemical behavior of pitch-based activated carbon fibers for electrochemical capacitors *Energy Convers. Manage.* **125** 347–52
[87] Ma C, Chen J, Fan Q, Guo J, Liu W, Cao E, Shi J and Song Y 2018 Preparation and one-step activation of nanoporous ultrafine carbon fibers derived from polyacrylonitrile/cellulose blend for used as supercapacitor electrode *J. Mater. Sci.* **53** 4527–39
[88] Jiang Q, Pang X, Geng S, Zhao Y, Wang X, Qin H, Liu B, Zhou J and Zhou T 2019 Simultaneous cross-linking and pore-forming electrospun carbon nanofibers towards high capacitive performance *Appl. Surf. Sci.* **479** 128–36
[89] Gaminian H, Montazer M, Bahi A, Karaaslan M and Ko F 2019 Capacitance performance boost of cellulose-derived carbon nanofibers via carbon and silver nanoparticles *Cellulose* **26** 2499–512
[90] Li Q, Deng L, Kim J K, Zhu Y Q, Holmes S M, Perez-Page M and Eichhorn S J 2017 Growth of carbon nanotubes on electrospun cellulose fibers for high performance supercapacitors *J. Electrochem. Soc.* **164** A3220
[91] Geim A K and Novoselov K S 2007 The rise of graphene *Nat. Mater.* **6** 183–91
[92] Huang Y, Liang J and Chen Y 2012 An overview of the applications of graphene based materials in supercapacitors *Small* **8** 1805–34
[93] Xia J, Chen F, Li J and Tao N 2009 Measurement of the quantum capacitance of graphene *Nat. Nanotechnol.* **4** 505–9

[94] Booth T J *et al* 2008 Macroscopic graphene membranes and their extraordinary stiffness *Nano Lett.* **8** 2442–6

[95] Lee C, Wei X, Kysar J W and Hone J 2008 Measurement of the elastic properties and intrinsic strength of monolayer graphene *Science* **321** 385–8

[96] Wu Z S, Ren W, Gao L, Zhao J, Chen Z, Liu B, Tang D, Yu B, Jiang C and Cheng H M 2009 Synthesis of graphene sheets with high electrical conductivity and good thermal stability by hydrogen arc discharge exfoliation *ACS Nano* **3** 411–7

[97] Ma T Y, Dai S, Jaroniec M and Qiao S Z 2014 Graphitic carbon nitride nanosheet–carbon nanotube three-dimensional porous composites as high-performance oxygen evolution electrocatalysts *Angew. Chem.* **126** 7409–13

[98] Leenaerts O, Partoens B and Peeters F M 2009 Adsorption of small molecules on graphene *Microelectron. J.* **40** 860–2

[99] Tan Y B and Lee J M 2013 Graphene for supercapacitor applications *J. Mater. Chem.* **1** 14814–43

[100] Sharmila T B, Sasi S, Suja N R, Beegum P S and Thachil E T 2019 A comparative investigation of aminosilane/ethylene diamine–functionalized graphene epoxy nanocomposites with commercial and chemically reduced graphene: static and dynamic mechanical properties *Emerg. Mater.* **2** 371–86

[101] Ju H F, Song W L and Fan L Z 2014 Rational design of graphene/porous carbon aerogels for high-performance flexible all-solid-state supercapacitors *J. Mater. Chem.* **2** 10895–903

[102] Becerril H A, Mao J, Liu Z, Stoltenberg R M, Bao Z and Chen Y 2008 Evaluation of solution-processed reduced graphene oxide films as transparent conductors *ACS Nano* **2** 463–70

[103] Pumera M 2010 Graphene-based nanomaterials and their electrochemistry *Chem. Soc. Rev.* **39** 4146–57

[104] Hummers W S Jr and Offeman R E 1958 Preparation of graphitic oxide *JACS* **80** 1339–9

[105] Kovtyukhova N I, Ollivier P J, Martin B R, Mallouk T E, Chizhik S A, Buzaneva E V and Gorchinskiy A D 1999 Layer-by-layer assembly of ultrathin composite films from micron-sized graphite oxide sheets and polycations *Chem. Mater.* **11** 771–78

[106] Park S and Ruoff R S 2009 Chemical methods for the production of graphenes *Nat. Nanotechnol.* **4** 217–24

[107] Fan Z J, Kai W, Yan J, Wei T, Zhi L J, Feng J, Ren Y M, Song L P and Wei F 2011 Facile synthesis of graphene nanosheets via Fe reduction of exfoliated graphite oxide *ACS Nano* **5** 191–8

[108] Li X *et al* 2009 Large-area synthesis of high-quality and uniform graphene films on copper foils *Science* **324** 1312–14

[109] Stoller M D, Park S, Zhu Y, An J and Ruoff R S 2008 Graphene-based ultracapacitors *Nano Lett.* **8** 3498–502

[110] Vivekchand S R C, Rout C S, Subrahmanyam K S, Govindaraj A and Rao C N R 2008 Graphene-based electrochemical supercapacitors *J. Chem. Sci.* **120** 9–13

[111] Wen Z, Wang X, Mao S, Bo Z, Kim H, Cui S, Lu G, Feng X and Chen J 2012 Crumpled nitrogen-doped graphene nanosheets with ultrahigh pore volume for high-performance supercapacitor *Adv. Mater.* **24** 5610–16

[112] Du X, Guo P, Song H and Chen X 2010 Graphene nanosheets as electrode material for electric double-layer capacitors *Electrochim. Acta* **55** 4812–19

[113] Wang Y, Shi Z, Huang Y, Ma Y, Wang C, Chen M and Chen Y 2009 Supercapacitor devices based on graphene materials *J. Phys. Chem.* **113** 13103–7

[114] Zhu Y, Murali S, Stoller M D, Velamakanni A, Piner R D and Ruoff R S 2010 Microwave assisted exfoliation and reduction of graphite oxide for ultracapacitors *Carbon* **48** 2118–22

[115] Purkait T, Singh G, Singh M, Kumar D and Dey R S 2017 Large area few-layer graphene with scalable preparation from waste biomass for high-performance supercapacitor *Sci. Rep.* **7** 1–14

[116] Zhao Y, Liu J, Hu Y, Cheng H, Hu C, Jiang C, Jiang L, Cao A and Qu L 2013 Highly compression-tolerant supercapacitor based on polypyrrole-mediated graphene foam electrodes *Adv. Mater.* **25** 591–5

[117] Xu Y, Lin Z, Zhong X, Huang X, Weiss N O, Huang Y and Duan X 2014 Holey graphene frameworks for highly efficient capacitive energy storage *Nat. Commun.* **5** 1–8

[118] Kim T, Jung G, Yoo S, Suh K S and Ruoff R S 2013 Activated graphene-based carbons as supercapacitor electrodes with macro-and mesopores *ACS Nano* **7** 6899–905

[119] Choi B G, Yang M, Hong W H, Choi J W and Huh Y S 2012 3D macroporous graphene frameworks for supercapacitors with high energy and power densities *ACS Nano* **6** 4020–28

[120] Jeong H M, Lee J W, Shin W H, Choi Y J, Shin H J, Kang J K and Choi J W 2011 Nitrogen-doped graphene for high-performance ultracapacitors and the importance of nitrogen-doped sites at basal planes *Nano Lett.* **11** 2472–77

[121] Wang H, Maiyalagan T and Wang X 2012 Review on recent progress in nitrogen-doped graphene: synthesis, characterization, and its potential applications *ACS Catal.* **2** 781–94

[122] Liu H, Liu Y and Zhu D 2011 Chemical doping of graphene *J. Mater. Chem.* **21** 3335–45

[123] Hu B, Wang K, Wu L, Yu S H, Antonietti M and Titirici M M 2010 Engineering carbon materials from the hydrothermal carbonization process of biomass *Adv. Mater.* **22** 813–28

[124] Liu Y *et al* 2016 Elemental superdoping of graphene and carbon nanotubes *Nat. Commun.* **7** 1–9

[125] Zhao Y, Hu C, Hu Y, Cheng H, Shi G and Qu L 2012 A versatile, ultralight, nitrogen-doped graphene framework *Angew. Chem. Int. Ed.* **51** 11371–5

[126] Han J, Zhang L L, Lee S, Oh J, Lee K S, Potts J R, Ji J, Zhao X, Ruoff R S and Park S 2013 Generation of B-doped graphene nanoplatelets using a solution process and their supercapacitor applications *ACS Nano* **7** 19–26

[127] Xu Y, Sheng K, Li C and Shi G 2010 Self-assembled graphene hydrogel via a one-step hydrothermal process *ACS Nano* **4** 4324–30

[128] Jung S M, Mafra D L, Lin C T, Jung H Y and Kong J 2015 Controlled porous structures of graphene aerogels and their effect on supercapacitor performance *Nanoscale* **7** 4386–93

[129] Zhang Q, Wang Y, Zhang B, Zhao K, He P and Huang B 2018 3D superelastic graphene aerogel-nanosheet hybrid hierarchical nanostructures as high-performance supercapacitor electrodes *Carbon* **127** 449–58

[130] Ye S, Feng J and Wu P 2013 Deposition of three-dimensional graphene aerogel on nickel foam as a binder-free supercapacitor electrode *ACS Appl. Mater. Interfaces* **5** 7122–29

[131] Xu P, Gao Q, Ma L, Li Z, Zhang H, Xiao H, Liang X, Zhang T, Tian X and Liu C 2019 A high surface area N-doped holey graphene aerogel with low charge transfer resistance as high performance electrode of non-flammable thermostable supercapacitors *Carbon* **149** 452–61

[132] Li J, Li X, Xiong D, Wang L and Li D 2019 Enhanced capacitance of boron-doped graphene aerogels for aqueous symmetric supercapacitors *Appl. Surf. Sci.* **475** 285–93

[133] Chen Y, Sun L, Lu Z, Liu Z, Jiang Y and Zhuo K 2019 Preparation of nitrogen and sulfur co-doped graphene aerogel with hierarchical porous structure using ionic liquid precursor for high-performance supercapacitor *Ionics* **25** 2781–9

[134] Wang C, Zhou Y, Sun L, Zhao Q, Zhang X, Wan P and Qiu J 2013 N/P-codoped thermally reduced graphene for high-performance supercapacitor applications *J. Phys. Chem.* **117** 14912–19

[135] Hassan M, Haque E, Reddy K R, Minett A I, Chen J and Gomes V G 2014 Edge-enriched graphene quantum dots for enhanced photo-luminescence and supercapacitance *Nanoscale* **6** 11988–94

[136] Cranford S W, Brommer D B and Buehler M J 2012 Extended graphynes: simple scaling laws for stiffness, strength and fracture *Nanoscale* **4** 7797–809

[137] Krishnamoorthy K, Thangavel S, Veetil J C, Raju N, Venugopal G and Kim S J 2016 Graphdiyne nanostructures as a new electrode material for electrochemical supercapacitors *Int. J. Hydrogen Energy* **41** 1672–8

[138] Zuo Z and Li Y 2019 Emerging electrochemical energy applications of graphdiyne *Joule* **3** 899–903

[139] Gao X, Liu H, Wang D and Zhang J 2019 Graphdiyne: synthesis, properties, and applications *Chem. Soc. Rev.* **48** 908–36

[140] Shang H, Zuo Z, Zheng H, Li K, Tu Z, Yi Y, Liu H, Li Y and Li Y 2018 N-doped graphdiyne for high-performance electrochemical electrodes *Nano Energy* **44** 144–54

Chapter 4

Comprehensive insight: electrode material selection for, and electrode performance in, pseudo and hybrid supercapacitors

To achieve a significant increase in the electrochemical performance of supercapacitors (SCs), enormous efforts have been made to replace electric double-layer supercapacitor (EDLC) active materials with pseudo-supercapacitive and composite materials. Metal oxides, metal sulfides, hydroxides, and conductive polymers are preferable classes of materials for pseudo-supercapacitors (PSCs). Hybrid SCs comprise a combination of EDLC and PSC materials for improved electrochemical performance. The simultaneous effects of a rapid reversible redox reaction and the formation of an electric double layer enhance the specific capacitance, rate capability, and cyclic performance of hybrid devices. Recently, for high-energy and -power applications, research into the development of hybrid SCs using a combination of a battery-type electrode and an SC electrode (capacitive or pseudocapacitive) has been explored. In this chapter, the most relevant materials for PSCs and hybrid SCs are discussed, with their main characteristics, merits, demerits, electrochemical performance, and recent trends.

4.1 Introduction

As discussed earlier, in chapter 2, pseudocapacitance arises due to reversible charge transfer faradaic processes at the interface between the electrode and the electrolyte. The electrochemical properties of PSCs are governed by the classes and combinations of materials chosen for their electrodes. Metal oxides and metal sulfides in binary, ternary, and composite forms are the most extensively investigated materials for PSCs. In this section, the characteristics of PSCs based on different forms of metal oxides and metal sulfides are discussed in detail.

doi:10.1088/978-0-7503-3103-6ch4

4.2 Metal oxides as electrode materials for supercapacitors

Metal oxides in different forms (doped, binary, and ternary) have been comprehensively reported as electrode materials for pseudosupercapacitors because of their high theoretical specific capacitance, low resistance, and tunable morphology. They can also provide more energy and power density than carbon materials, as they store energy by both faradaic and non-faradaic processes [1–4].

The important features of the metal oxides used for SC applications include:

- easily tailored electronic properties
- synthesis that takes place by several cost-effective methods using abundant resources.
- the ability to be designed using various nanoarchitectural morphologies (produced by optimizing the synthesis conditions), facilitating surface areas with increased electrochemical activity
- the ability to be used in crystalline, amorphous, or hydrated form
- superior electrochemical stability (lack of structural deformation)
- the ability to be synthesized with various other materials to form nanocomposites

The selection of a metal oxide for a PSC depends on:

1. Electric conductivity
2. The oxidation states of the metal
3. A structure that facilities intercalation into the oxide lattice

Figure 4.1 shows a schematic representation of various factors that affect the electrochemical performance of metal-oxide-based SCs. To date, various metal

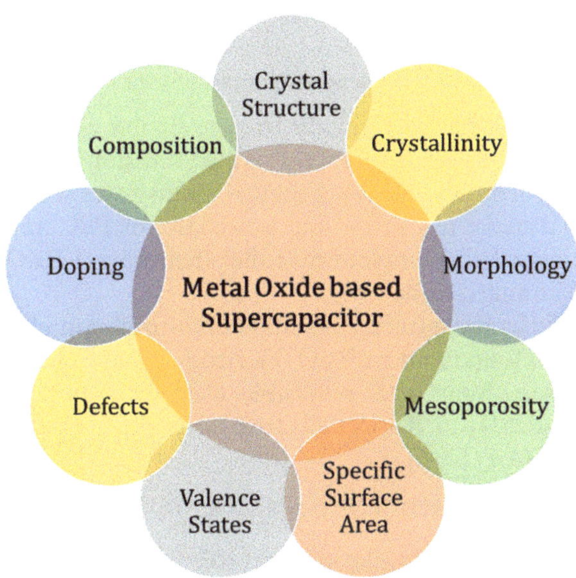

Figure 4.1. Schematic representation of the factors affecting the electrochemical performance of metal-oxide-based SCs.

Table 4.1. Metal oxides and their estimated theoretical capacitances.

Metal oxide	Theoretical capacitance ($F\ g^{-1}$)
RuO_2	~1400–2200
MnO_2	~1370
NiO	~2500
CoO	> 4200
CuO	~ 1800
V_2O_5	~2120

oxides, such as ruthenium dioxide (RuO_2), manganese oxide (MnO_2), nickel oxide (NiO), iridium oxide (IrO_2), cobalt oxide (CoO), vanadium oxide (V_2O_5, VO_2), iron oxide (Fe_2O_3, Fe_3O_4), copper oxide (CuO), titanium oxide (TiO_2), tin oxide (SnO_2), etc, have been widely investigated as electrode materials for PSCs. The theoretical capacitances of some commonly used metal oxides are listed in table 4.1.

Various cost-effective methods have been adopted for the synthesis of metal oxides, including the wet chemical, sol–gel, hydrothermal, and electrodeposition methods, template-assisted synthesis, anodic deposition, solid-state thermochemistry, mechano-milling, etc. By optimizing the synthesis parameters, the properties of materials, such as their specific surface area, porosity, and pore size distribution and their physicochemical, electrical, thermal, and microstructural properties can be tailored [3–5]. Amongst the metal oxides, RuO_2 and MnO_2 have been widely explored as electrode materials and have shown promising pseudocapacitive performance.

4.2.1 Performance of RuO_2-based pseudosupercapacitors

The pseudocapacitance of RuO_2 originates from reversible electron transfer and the electroadsorption of protons on its surface. It is also stable in three distinct oxidation states within a 1.2 V potential window. Figure 4.2 illustrates some of the important features of RuO_2 that make it a promising candidate for PSC applications.

The electrochemical performance of SCs that incorporate RuO_2 is governed by various factors, as summarized below.

RuO_x has shown different specific capacitances for different electrolytes. In some electrolytes, it has been found that an RuO_2 electrode behaves like that of an EDLC. In an acidic medium, along with reversible electron transfer, Ru changes its oxidation state from Ru(II) to Ru(IV) electroadsorption of protons on its surface, as shown by equation (4.1) [2, 6],

$$RuO_2 + xH^+ + xe^- \leftrightarrow RuO_{2-x}(OH)_x. \tag{4.1}$$

The electrolyte concentration also plays a vital role in governing its supercapacitive performance. The optimum ion concentration in the electrolyte fulfils the needs of both the electrical double layer and the faradaic reactions.

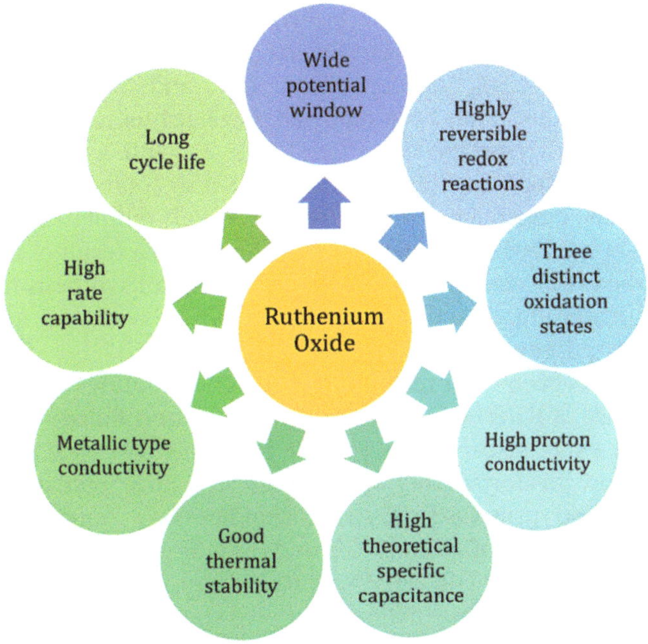

Figure 4.2. Schematic illustration of the features of RuO_2 as an electrode material for SCs.

Its performance is also sensitive to the RuO_x content in the electrode, the crystallinity, and the type of electrolyte used. RuO_2 can be crystalline or amorphous, while crystalline RuO_2 has two forms, anhydrous (rutile, RuO_2) and hydrous ruthenium oxide ($RuO_2{\cdot}xH_2O$). Compared with rutile RuO_2, hydrous $RuO_2{\cdot}xH_2O$ shows a much better specific capacitance. $RuO_2{\cdot}xH_2O$ has hydrous mesopores, which can be expected to boost the diffusion of cations inside the electrode, hence leading to an increase in its specific surface area and specific capacitance. The hydrous regions in $RuO_2{\cdot}xH_2O$ behave as good proton donors, helping to improve the ion diffusion kinetic processes [6–8]. It has also been reported that $RuO_2{\cdot}xH_2O$, with its amorphous nature, performed better than the crystalline form, because in amorphous materials, redox reactions on the surface are accompanied by those inside the bulk of the electrode. In addition, compared to the crystalline form, amorphous RuO_2 exhibits higher specific capacitance, due to the availability of greater active electrochemical sites where the redox reaction takes place [9, 10].

An ordered porous hierarchical network also helps to improve electrochemical performance due to its superior ion diffusion process. RuO_2 synthesized at a smaller size and with different nanostructures exhibited superior specific capacitance as a result of the enhanced surface area and conductivity and the easy path for diffusion. Nanosized $RuO_2{\cdot}xH_2O$ particles are expected to shorten the diffusion path and help protonic conduction. A smaller size and an increase in the surface-to-volume ratio resulted in an enhancement of the electroactive sites, which further led to higher energy density and capacitance. The particle size was also controlled by adjusting the preparative parameters. For instance, in colloidal synthesis, the particle size is controlled by the amount of sodium bicarbonate ($NaHCO_3$) and the reaction time [11]. In electroless deposition, the desired particle size is achieved by varying the pH of the reaction bath, the deposition time, and the temperature [12]. In template-assisted methods, it has been reported that hard-template-assisted synthesis leads to a relatively low specific capacity [13], even with a regular and ordered porous network. This fact is attributed to the low water content of $RuO_2{\cdot}xH_2O$ due to the high pyrolysis temperature used. Conversely, soft-template-assisted synthesis leads to a threefold increase in the specific capacitance, as a result of increased water content, due to the lower temperature used [14–16]. Polyol synthesis has been found to be the most promising option for template-assisted synthesis, as it leads to a uniform size distribution and a sufficient water content [17].

Hydrous RuO_2 has been explored in different nanostructures for SC applications [18–21]. For instance, a RuO_2 thin film with a porous morphology prepared on a Ti substrate exhibited a specific capacitance of 788 F g^{-1} in 0.5 M H_2SO_4 electrolyte [18]. Furthermore, hollow fusiform $RuO_2{\cdot}xH_2O$ nanostructures prepared via a hydrothermal approach and using α-Fe_2O_3 nanospindles as a template demonstrated good specific capacitances of about 772 and 671 F g^{-1} at current densities of 0.5 and 4 A g^{-1}, respectively [19]. Hydrous RuO_2 nanotubes developed directly on a Ti substrate via a facile template method exhibited specific capacitances of 840 F g^{-1} and 745 F g^{-1} at current densities of 2 A g^{-1} and 32 A g^{-1}, respectively, with a good capacitive retention of 88.7% in 1 M H_2SO_4 electrolyte [20].

Kim *et al* prepared nanosheet-assembled 3D nanoflowers of ruthenium oxide by a microwave hydrothermal method, which exhibited specific capacitances of 545.2 and 498.2 F g^{-1} at current densities of 0.5 and 50 A g^{-1}, respectively, indicating a decrease of only 8.6% in the specific capacitance even at a high current density. The cyclic performance of a nanoflower electrode was measured at constant scan rate of 50 mV s^{-1}, which showed capacitance retentions of about 89.8 and 83.4% after 2000 and 5000 cycles, respectively. The enhanced performance was attributed to the hierarchical porous 3D nature of nanoflowers, which contributed to fast kinetic processes [22].

Thus, in different nanostructures, RuO_2 exhibited high specific capacitance and remarkable rate capability due to a well-crystallized porous structured network and the unique morphology of the prepared nanomaterials. Even though ruthenium oxide has exceptional properties and high theoretical capacitance, it is not considered for practical applications because of its toxic nature, high cost, and low operating voltage window.

Several efforts have been made to synthesize doped ruthenium oxide and mixed-oxide composites [23–25] to reduce the Ru content without suppressing performance. RuO_2 mixed with other abundant and low-cost metal oxides, such as VO_2, SnO_2, Co_3O_4, TiO_2, MnO_2, MoO_3, WO_3, and NiO, has been reported [12, 14, 26–31]. For instance, it has been reported that the doping of RuO_2 with vanadium (V) increases its electrochemical stability and the utilization of the Ru species and also extends the potential window [23, 25]. Additionally, vanadium doping reduces the particle size of RuO_2, leading to a high specific surface area [24]. Li *et al* reported a TiO_2-SnO_2 doped RuO_2 composite electrode that exhibited a specific capacitance of 571 F g^{-1} with

35 wt% TiO_2-SnO_2 [31]. Lei *et al* reported a RuO_2/MnO_2 composite electrode with a specific capacitance of 438 F g^{-1} for an electrode with a 60% mass ratio of manganese oxide and a capacitance retention of 92.5% over 300 cycles in a 38% mass ratio H_2SO_4 electrolyte [29]. An SC with a $Co_3O_4/RuO_2{\cdot}xH_2O$ composite electrode showed a high specific capacitance of 642 F g^{-1} for a composite with a Co to Ru ratio of 1:1; it also exhibited good cycle stability [14]. Wang *et al* prepared RuO_2/TiO_2 nanotube composites by loading various amounts of RuO_2 onto pre-synthesized TiO_2 nanotubes. An SC with a high-aspect-ratio composite with 4 wt% of RuO_2 exhibited a very high capacitance of 1263 F g^{-1} in a symmetric two-electrode cell [32]. The enhanced performance was attributed to easily accessible RuO_2 electrochemical active sites for electrochemical kinetics, and the 3D nanotube network supported a fast and easy ion diffusion process.

In all the above studies, attempts have been made to reduce the amount of RuO_2 in order to decrease the cost and toxicity, but those resulted in poor performance due to a significant change in electrode conductivity. The composites had high interfacial charge transfer resistances due to their heterogeneous structures and nanoparticle aggregation, which affected conductivity and hence resulted in poor electrochemical performance. Some of the promising results for SCs based on RuO_2 are summarized in table 4.2.

4.2.2 Performance of MnO_2-based pseudosupercapacitors

To overcome the limitations of RuO_2-based materials, several abundant, nontoxic, and less expensive oxides have been widely investigated. After RuO_2, MnO_2 is the most extensively explored metal oxide for SC applications because of its high theoretical specific capacity (1370 F g^{-1}) [33–36].

To date, several studies have reported the superior performance of MnO_2 and MnO_2-based electrodes. MnO_2 can be synthesized in different nanoarchitectural forms using various techniques, such as the wet chemical, electrochemical, hydrothermal, solid-state, and mechanomilling methods, etc [37–45].

Pseudocapacitance in manganese oxides

Pseudocapacitance in manganese oxides is attributed to reversible transitions between Mn(II)/Mn(III) and Mn(III)/Mn(IV) due to redox reactions within the potential window of the electrolyte. Such a transition is facilitated in two ways [33]:

1. Reversible transitions occur due to the exchange of protons and/or cations with the electrolyte, i.e. the intercalation of protons (H^+) during the charging process followed by de-intercalation during discharging.
    - Intercalation:

$$MnO_2 + [H^+] \rightarrow e^- + MnOOH \tag{4.2}$$

    - De-intercalation:

$$MnOOH + e^- \rightarrow MnO_2 + [H^+] \tag{4.3}$$

Table 4.2. Electrochemical performance of PSCs based on metal oxides, mixed metal oxides, and their nanostructures.

Working electrode	Electrolyte	Potential window (V)	Specific capacitance (F g^{-1}) at (@) current density (A g^{-1})/scan rate (mV s^{-1})	Power density (W kg^{-1})	Energy density (Wh kg^{-1})	Retention % (cycles) at (@) current density (A g^{-1})/scan rate (mV s^{-1})	References
Hollow fusiform RuO_2. xH_2O	1 M H_2SO_4	0 to 1	772@0.5 and 671@4				[19]
Hydrous RuO_2 nanotubes on a Ti substrate	1 M H_2SO_4	0.1 to 0.8	840@ 2 and 745@32			60% (1400) @16	[20]
3D RuO_2 nanoflowers	1 M H_2SO_4	0 to 1	545.2@0.5 and 498.2@50			89.8% (2000) and 83.4% (5000)/50	[22]
α-MnO_2 (amorphous)	0.25 M Na_2SO_4	0 to 0.8	200@1				[49]
γ-MnO_2 nanosheets (NSs)	1 M Na_2SO_4	0 to 1	162@20			92% (5000)	[51]
2D β-MnO_2 nanowire network	1 M Na_2SO_4	0 to 0.8	453@0.5	8800		50% (1800)	[52]
Layered structured δ-MnO_2	0.5 M Na_2SO_4	−0.2 to 1.2	344/5				[53]
3D spinel λ -MnO_2	0.5 M K_2SO_4	0 to 1	241/5			>100% (5000)	[56]
CuO@NiO_2 core–shell nanostructure	1 M Na_2SO_4	−0.2 to 0.8	374@0.25			86.9% (3000)	[60]
Porous WO_3@ CuO composites	6 M KOH	0 to 0.5	284@1			85.2% (1500)	[62]
$NiCo_2O_4$ nanotubes	2 M KOH	0 to 0.4	1647@1			93.6% (3000)	[69]
$NiCo_2O_4$ nanoflowers	2 M KOH	0 to 0.6	1191.2@1 and 755.2@10			63.4%@10	[70]

(Continued)

Table 4.2. *(Continued)*

Working electrode	Electrolyte	Potential window (V)	Specific capacitance (F g^{-1}) at (@) current density (A g^{-1})/scan rate (mV s^{-1})	Power density (W kg^{-1})	Energy density (Wh kg^{-1})	Retention % (cycles) at (@) current density (A g^{-1})/scan rate (mV s^{-1})	References
$MnCo_2O_4$ (MCO) nanosheet arrays	2 M KOH	−0.1 to 0.4	290/1	5200	10.04		[71]
$NiMoO_4$ nanowires	2 M KOH	−0.2 to 1	1587@5 mA cm^{-2} and 951@30 mA cm^{-2}	16 000	70.7	76.9% (4000)	[72]
$MnMoO_4/CoMoO_4$ nanowire heterostructures	2 M NaOH	−0.6 to 0.4	204.1@ 0.5187.1@1 and 134.7@3		28.4	98% (1000)@3	[73]
$NiCo_2O_4$ nanosheet-decorated $MnCo_2O_4$ nanoflakes	3 M KOH	0 to 0.6	1152@1	248.9	40	95.38% (3000)	[74]
$ZnO@MnCo_2O_4$	3 M KOH	−0.2 to 0.6	631.2@1	406	56.10	93.5% (3000)	[75]
$ZnCo_2O_4@MnCo_2O_4$	6 M KOH	0 to 0.5	254@1	750	19.5	98.5% (2000)	[76]
$MnCo_2O_4@MnO_2$	3 M KOH	0 to 0.7	858@1	252	36	91% (5000)	[77]
$MnMoO_4/NiWO_4$	2 M KOH	0 to 0.4	598@1			82% (5000)	[78]
$ZnMn_2O_4@ZnFe_2O_4$	3 M KOH	0 to 5	1024.66@10 mA cm^{-2}			95.8% (3000)	[79]
$Co_3O_4@NiCo_2O_4$	2 M KOH	0 to 0.5	672@1			97.1% (3000)	[80]

2. The electrolyte cations of alkali metals are adsorbed on the surface; the second mechanism explains the charge storage in terms of the surface adsorption of alkali metal cations (M^+), such as K^+, Li^+, etc, on the surface of MnO_2, which can be expressed as follows:
   - At the surface:

$$MnO_2 + [M^+] + e^- \rightarrow [MnOO^-M^+]. \tag{4.4}$$

It has been reported that not all Mn ions change their oxidation state in the case of bulk MnO_2; such a partial contribution limits the rate of protonic diffusion, and hence hinders the pseudocapacitive performance of bulk MnO_2 [42].

Factors affecting pseudocapacitance in manganese oxides

Crystal structure, crystallinity, and morphology

MnO_2 has six different crystallographic phases. The atomic structure consists of one manganese atom surrounded by six oxygen atoms to form an octahedron, MnO_6. The Mn^{4+} ions at the center and the O^{2-} ions at the corners have spin-polarized $3d^3$ and spin-unpolarized $2p^6$ configurations, respectively. The O^{2-} ions also share the edges of the MnO_6 octahedral units, whose compact or dense packing allows for six different forms of MnO_2, namely α-MnO_2, β-MnO_2, R-MnO_2, γ-MnO_2, δ-MnO_2, and λ-MnO_2. Among these, the first four have tunnel structures, while δ-MnO_2 and λ-MnO_2 have layered and spinel structures, respectively [46–48]. The electrical conductivity, ionic diffusivity, porous network formation, and pseudocapacity of MnO_2 are greatly influenced by its crystal structure.

The morphology of MnO_2 is one of the crucial factors that affect its supercapacitive performance, as it decides the active surface area available for electrochemical processes. Several morphologies of MnO_2, such as nanosheets, nanowires, nanorods, nanobelts, and nanoflowers have been reported for SC applications [37, 44,45].

For instance, Zhang *et al* studied the supercapacitive performance of manganese oxides with different crystallographic phases in a sodium sulphate (Na_2SO_4) electrolyte. The authors reported that the specific capacitance values decreased with increasing crystallinity, because increasing crystallinity limits the protonation process. The specific capacitance increased in the following order: $\beta < \gamma < \delta < \alpha$-$MnO_2$ [49]. The study reported a specific capacitance of 200 F g^{-1} at 1 A g^{-1} in 0.25 M Na_2SO_4 electrolyte, which is much higher than that of single-crystal α-MnO_2 [50].

Park *et al* synthesized uniform γ-MnO_2 nanosheets on highly conductive one-dimensional backbone arrays, which exhibited a specific capacitance of 162 F g^{-1} at 20 A g^{-1} and a good capacitance retention of about 92% over 5000 cycles in 1 M Na_2SO_4 aqueous electrolyte [51].

Wei *et al* reported a 2D β-MnO_2 nanowire network with a specific capacitance of 453 F g^{-1} at 0.5 A g^{-1} in 1 M Na_2SO_4 aqueous electrolyte [52]. Zolfaghari *et al* synthesized layer-structured δ-MnO_2 by a sonochemical method, which exhibited a specific capacitance of 344 F g^{-1} at 5 mV s^{-1} in 0.5 M Na_2SO_4 electrolyte [53]. It has been reported that the layered structure of δ-MnO_2 facilitates the intercalation of alkali metal cations [54, 55]. Ghodbane *et al* prepared spinel (λ -MnO_2), which exhibited a specific capacitance of 241 F g^{-1} at 5 mV s^{-1} in 0.5 M K_2SO_4 electrolyte [56]. Li *et al* synthesized MnO_2 nanotubes by a hydrothermal method, which exhibited a specific capacitance of 411.9 F g^{-1} at 0.25 A g^{-1} and a good capacitance retention of 90.4% over 3000 cycles in 1 M Na_2SO_4 electrolyte [57]. Wu reported MnO_2 nanowires with a specific capacitance of 350 F g^{-1} at 0.1 mA cm^{-2} in 0.1 M Na_2SO_4 electrolyte [58].

Similarly to ruthenium oxide, MnO_2 in its hydrous form, $MnO_2{\cdot}nH_2O$, has also been extensively studied for SC applications. Amorphous hydrous manganese oxide ($MnO_2{\cdot}nH_2O$) deposited anodically on a graphite substrate exhibited a specific capacitance of about 265 to 320 F g^{-1} in 0.1 M Na_2SO_4 electrolyte [59].

Thus, nanostructured, layered, and amorphous structures showed better performance in MnO_2-based PSC due to their porous networks and easy processes of ion diffusion and charge conduction through either a redox reaction or intercalation. However, its comparatively low surface area, partial dissolution in electrolytes, poor ionic conductivity (10^{-13} S cm^{-1}), poor electrical conductivity (10^{-5}–10^{-6} S cm^{-1}), and expensive synthesis reagents are major constraints for the practical application of MnO_2.

Many other metal oxide materials, such as WO_3, NiO, IrO_2, CoO, Co_3O_4, V_2O_5, Fe_2O_3, Fe_3O_4, CuO, TiO_2, SnO_2 have also been extensively explored as electrodes for SC applications. However, none of them has achieved its theoretical estimated specific capacitance in its intrinsic form. The specific capacitance of metal oxides in their intrinsic forms lies in the range of 20 to 500 F g^{-1} at various current densities. The optimization of one parameter causes problems for another; for example, a porous morphological structure increases the specific surface area, but it causes a decrease in the electrical conductivity. In some cases, a higher specific capacitance has been obtained, but only in low potential ranges, which obstructs practical use. Some of the prominent results for SCs based on metal oxides are summarized in table 4.2

4.2.3 Performance of pseudosupercapacitors based on mixed metal oxides

To further improve specific capacitance, certain changes in the intrinsic forms of metal oxides have been considered to be effective. Mixed metal oxides can be obtained either by doping with metal ions or by preparing composites of metal oxides. It seems that the incorporation of suitable metals (with low and high oxidation states) into a host metal oxide can yield a much higher specific capacitance than that of the host material. Multiple oxidation states allow multiple redox reactions to take place during the electrochemical process, leading to higher supercapacitive performance. The synchronization of properties overcomes obstacles and leads to enhanced SC performance. For instance, CuO@NiO and CuO@MnO_2 nanocomposites showed much better performance than that of CuO, due to an enhancement in electrical conductivity [60, 61]. Tian *et al* prepared a WO_3@CuO composite, which showed a specific capacitance of 284 F g^{-1} at 1 A g^{-1} (which is much higher than that of a pure CuO electrode (48 F g^{-1})) and a capacitance retention of about 85.2% over 1500 cycles. The enhanced performance of the composite was attributed to the small radius, high intrinsic density (> 7 g cm^{-3}), and high electronic conductivity (10^{-6} S cm^{-1}) of WO_3 [62].

Ternary metal oxides (TMOs) are another form of mixed metal oxide; their general formula is $A_xB_yO_z$, where A and B are metal elements with different oxidation states. According to the ratio of A:B:O, TMOs are categorized into three different groups: AB_2O_4, $ABO_{2/3/4}$, and $A_3B_2O_8$. Supercapacitor studies based on TMOs have mainly focused on designing them using different nanostructures produced by various methods in order to improve their electrical conductivity, specific surface area, ion diffusion kinetics, etc. TMOs such as $NiCo_2O_4$, $ZnCo_2O_4$, Zn_2SnO_4, $Co_3V_2O_8$, $MnCo_2O_4$, $MnFe_2O_4$, and $LiMn_2O_4$ have been widely explored. In particular, TMO materials in 1D to 3D nanostructures have been extensively explored due to their higher electrical conductivity and good capacitance retention over thousands of cycles. Porous nanostructures guarantee an increased effective contact area between the electrolyte and the active materials; therefore an increased number of electrochemical active sites become available for electrochemical processes along with easy pathways for charge transport in PSCs [63–68].

For example, Li *et al* studied the supercapacitive performance of $NiCo_2O_4$ nanotubes, which showed a specific capacitance of 1647 F g^{-1} at 1 A g^{-1} with a capacity retention of 77.3% at 25 A g^{-1}. Their superior performance was attributed to their porous architecture and hollow interiors, which enhanced electrochemical activity [69]. An *et al* synthesized patterned $NiCo_2O_4$ nanoflowers assembled by the radial growth of nanosheets from a particular center, which revealed a higher specific capacitance of 1191.2 F g^{-1} at 1 A g^{-1} with a capacitance retention of 77.75% after 1200 cycles [70]. Sahoo *et al* prepared $MnCo_2O_4$ (MCO) nanosheet arrays on an ITO substrate for SC applications. Their study revealed a specific capacitance, an energy density, and a power density of 290 F g^{-1}, 10.04 Wh kg^{-1}, and 5.2 kW kg^{-1}, respectively, at 1 mV s^{-1} in 2 M KOH electrolyte [71]. Guo *et al* radially prepared $NiMoO_4$ nanowires on a carbon cloth, which showed specific capacitances of 1.27 F cm^{-2} (1587 F g^{-1}) at 5 mA cm^{-2} and 0.76 F cm^{-2} (951 F g^{-1}) at 30 mA cm^{-2}. The device showed a maximum voltage of 1.7 V, an energy density of 70.7 Wh kg^{-1}, a power density of 16 000 W kg^{-1} at 14.1 Wh kg^{-1}, and a retention of 76.9% after 4000 cycles [72].

Recently, some studies of composite TMOs have also been carried out to obtain further improvements in specific capacitance and stability. Mai *et al* prepared hierarchical $MnMoO_4/CoMoO_4$ nanowire heterostructures, which exhibited specific capacitances of 204.1 and 187.1 F g^{-1} at 0.5 and 1 A g^{-1}, respectively. The performance obtained was higher than that of an electrode made using $CoMoO_4$ nanowires [73].

Lee *et al* synthesized $NiCo_2O_4$ (NCO) nanosheet-decorated $MnCo_2O_4$ (MCO) nanoflakes on nickel foam. The electrochemical performance of the NCO, MCO, and MCO @NCO composite electrode is shown in figure 4.3. The cyclic voltammogram (CV) curves (figure 4.3(a)) of MCO, NCO, and MCO @NCO at a scan rate of 20 mV s^{-1} with a potential range 0 to 0.6 V show a pair of redox peaks, representing the occurrence of specific capacitance due to faradaic redox reactions. The non-rectangular shape of the CV curves substantiates the pseudocapacitive behavior of the electrode. The large integrated area and the presence of strong redox peaks in the MCO @NCO composite curves confirm its greater electrochemical performance.

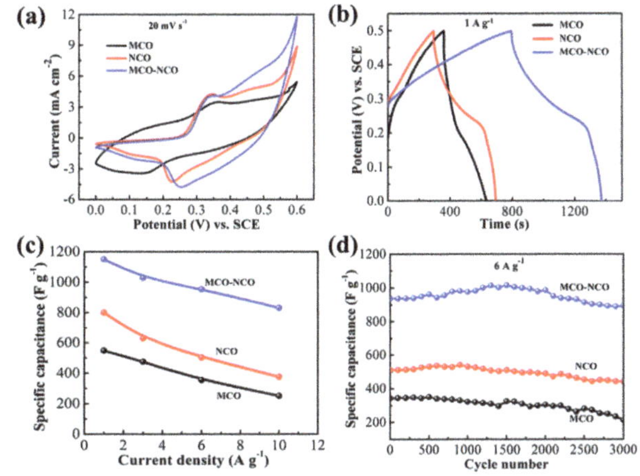

Figure 4.3. Electrochemical performance of NCO, MCO, and NCO@MCO composite electrodes: (a) CV curves at a scan rate of 20 mV s^{-1}, (b) GCD curves at 1 A g^{-1}, (c) the variation of specific capacitance for different current densities, and (d) the cycling performance at 6 A g^{-1}. Reproduced from [74] with permission of The Royal Society of Chemisty.

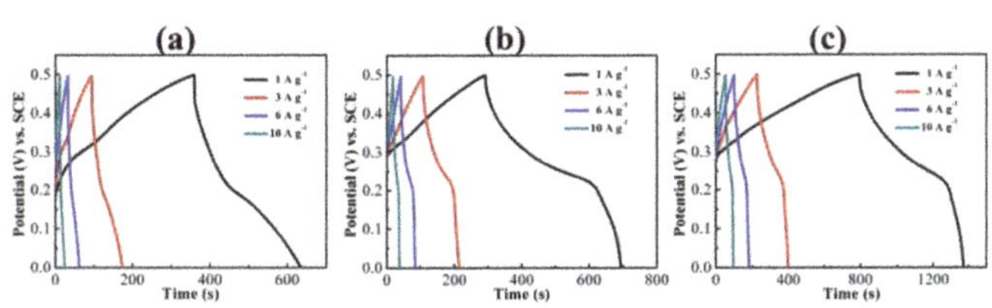

Figure 4.4. GCD plots of: (a) NCO, (b) MCO, and (c) NCO@MCO composite electrodes at different current densities. Reproduced from [74] with permission of The Royal Society of Chemisty.

The GCD curves in figure 4.3(b) at 1 A g^{-1} also indicate the pseudocapacitor-type behavior of the MCO, NCO, and MCO@NCO electrodes. Figure 4.4 shows GCD plots of the MCO, NCO, and MCO@NCO electrodes at current densities of 1, 3, 6, and 10 A g^{-1}. The small IR drop that can be observed for all samples may be due to internal resistance and contact resistance at the interface. However, the said loss is minimized in the case of MCO@NCO, due to the efficient penetration of electrolyte ions into the electrode material. The variation of the specific capacitance calculated using the GCD curves is represented in figure 4.3(c) for different current densities. The specific capacitances of MCO@NCO calculated using the GCD curves have values of 1152, 1029.6, 955.2, and 832 F g^{-1} at current densities of 1, 3, 6, and 10 A g^{-1}, respectively, which are much greater than those of singular MCO and NCO. Figure 3(d) depicts the cyclic performance of MCO, NCO, and MCO@NCO electrodes at a current density of 6 A g^{-1}. The MCO@NCO electrode showed an excellent retention of 95.38% after 3000 cycles. The measured energy density (40 Wh kg^{-1}) and power density (248.9 W kg^{-1}) of NCO@MCO at a current density of 1 A g^{-1} were found to be much higher than those of MCO (19 Wh kg^{-1} at 249.6 W kg^{-1}) and NCO (27.8 Wh kg^{-1} at 249.7 W kg^{-1}) [74]. The electrochemical properties given in some recent reports for $ZnO@MnCo_2O_4$, $ZnCo_2O_4@MnCo_2O_4$, $MnCo_2O_4@MnO_2$, $MnMoO_4@NiWO_4$, and $ZnMn_2O_4@ZnFe_2O_4$ are tabulated in table 4.2 [75–80].

4.3 Metal sulfides as electrode materials for supercapacitors

Metal sulfides (MSs) are another well-explored class of material that has recently been investigated for high-energy SCs. They offer larger crystal lattice distances that are suitable for the intercalation/de-intercalation processes and which have adaptable redox chemistry, high electrical conductivity, and tunable electronic charge transport properties and good electrochemical, thermal, and mechanical stability [81–83]. The electrochemical properties of MSs are much better than those of MOs and carbon-based materials. The lower electronegativity of sulfur compared to that of oxygen facilitates easier charge transport through the structures of MSs [84].

Charges are stored in MSs by reversible faradaic redox reactions using a two-step mechanism. In the first step, when MSs react with OH^-, they are converted to sulfo-hydroxide MSOH species, which are further changed into sulfoxide, MSO, in the second step [85].

$$\mathrm{MS} + \mathrm{OH}^- \leftrightarrow \mathrm{MSOH} + e^- \tag{4.5}$$

$$\mathrm{MSOH} + \mathrm{OH}^- \leftrightarrow \mathrm{MSO} + \mathrm{H_2O} + e^-. \tag{4.6}$$

For example: the energy stored in a CoS_x-based SC is expressed by the faradaic reaction between CoS_x and OH^-. The electron transition takes place between a Co^{2+}/Co^{3+} redox couple [86–88]. The reactions are:

$$\mathrm{CoS}_x + \mathrm{OH}^- \leftrightarrow \mathrm{CoS}_x\mathrm{OH} + e^- \tag{4.7}$$

$$\mathrm{CoS}_x\mathrm{OH} + \mathrm{OH}^- \leftrightarrow \mathrm{CoS}_x\mathrm{O} + \mathrm{H_2O} + e^-. \tag{4.8}$$

According to their lattice structure, MSs are classified into two categories: layered and non-layered structures. Layered metal sulfides have graphene-like structures and are composed of three atomic layers bound by covalent bonds. The individual atomic layers are held together by van der Waals forces. Layered MSs offer a channeled structure for ion intercalation and also provide an increased number of electrochemical active sites for charge storage/transportation within their structures. The most commonly employed materials are MoS_2, WS_2, SnS_2, VS_2, etc. However, most research has been focused on optimizing the electrical conductivity and surface area [81–83], even though the availability of sufficient volume for the intercalation of charges in the electrode structure is also equally important. Non-layered MS structures have adequate surface active sites due to the existence of surface dangling bonds, and hence are also capable electrode materials for electrochemical charge. The most commonly reported non-layered MSs are FeS_2, NiS_2, CoS_2, MnS, and GeS [81–83, 89].

Recent progress in MSs has been focused on the synthesis of nanostructured porous networks to improve the specific surface area along with the structural, morphological, electronic, and electrochemical properties. Several attempts have been made (summarized in table 4.3) to improve the specific capacitance of MSs or

Table 4.3. Electrochemical performance of PSCs based on metal sulfides, mixed metal sulfides, and their different nanostructures.

Working electrode	Electrolyte	Potential window (V)	Specific capacitance (F g^{-1}) at (@) current density (A g^{-1})/scan rate (mV s^{-1})	Power density (W kg^{-1})	Energy density (Wh kg^{-1})	Retention % (cycles) at (@) current density (A g^{-1})/scan rate (mV s^{-1})	References
NiS	2 M KOH	0.1 to 0.6	964@1	250	16.5	100% (2000)	[90]
3D Ni_3S_2	1 M KOH	0 to 0.6	626.1@5				[91]
Ni_3S_2	1 M NaOH	0 to 0.8	1000@50	17 000	202		[92]
CoS	6 M KOH	0 to 0.4	570@1	700.12	15.58	97.9% (2000)	[93]
Hollow CoS_2	2 M KOH	−0.2 to 0.4	936@1			83% (5000)@5	[94]
Hierarchical CoS_2	2 M KOH	−0.2 to 0.6	718.7@1			66.3% (1000)	[95]
Pyrite (FeS_2)	1 M Na_2SO_4	−1 to 0	308.3/5 and 317.8@3			81.1% (1000)	[96]
FeS_2	1 M $LiClO_4$	0 to 1.2	193@2.4 mA cm^{-2}	726	2.56	99% (1000)	[97]
WS_2	3 M KOH	0 to 0.5	1439.5@5 mA cm^{-2}			77.4% (3000)	[98]
MoS_2	3 M KOH	0 to 0.5	1531.2@5 mA cm^{-2}			81.6% (3000)	[98]
VS_2	PVA-$LiClO_4$	−0.6 to 0.2	349/2	6.2	25.9	89% (6000)	[99]
SnS_2	1 M KOH	−0.2 to −0.8	201@0.1	1200		88% (1000)	[100]
Bi_2S_3	6 M KOH	−0.1 to −0.9	550@0.5	550	87	87% (500)	[101]
γ-MnS	0.5 M Na_2SO_4	−0.1 to 0.4	378@0.2			90% (4000)	[102]
CuS hollow nanospheres@Ni foam	6 M KOH	−0.2 to 0.4	948@1	300	47.4	95.5% (1000)	[103]
CuS nanowalls@Ni-foam	2 M KOH	−0.3 to 0.6	1124@15 mA cm^{-2}			80.5% (2000)	[104]
Nickel cobalt sulfide	1 M KOH	0 to 1.75	1418@5	1800	60	63.2% (50 000)	[107]
$CoNi_2S_4$ nanoparticles	3 M KOH	0 to 0.5	1169@1	446.6	46.9	99% (2000)	[108]
$CoNi_2S_4$ nanosheets	2 M KOH	−0.1 to 0.6	2906@5 mA cm^{-2}	409	33.9	78% (3000)	[109]
Mesoporous $CuCo_2S_4$	2 M KOH	0.0 to 0.5	752@2			98.1% (5000)	[110]

$CuCo_2S_4$ hollow nanoneedles@Ni foam	3 M KOH	−0.1 to −0.55	2163@6 mA cm^{-2}	800	44.1	98.7% (1000)	[111]
$CuCo_2S_4$	Polysulfide	−0.25 to 0.4	5030@20			79.5% (2000)	[112]
$CoMoS_4$	1 M KOH	0.0 to 0.7	661@1	400	27.2	86% (10 000)	[113]
Flower-like manganese cobalt oxysulfide	1 M KOH	−0.1 to −0.65	490@2			86.5% (3000)	[114]
Manganese cobalt sulfide nanosheets	2 M KOH	−0.1 to −0.4	2421 @1	5650	83.83		[115]
NiV_2S_4 nanosheet on a 3D Ni framework	6 M KOH	0.0 to −0.5	639@2 mA cm^{-2}	240	45.1	90.7% (2000)	[116]
$SnNi_2S_4$	1 M KOH	−0.2 to 0.8	1484@2	700.3	64.38		[117]
NiS/Ni_3S_4 composite	2 M KOH	−0.2 to 0.6	700@10			65.8% (5000)	[118]
Ni_3S_2@MoS_2 core–shell	2 M KOH	−0.2 to 0.6	848@5			91% (2000)	[119]
MoS_2-coated CoS_2 composite	2 M KOH.	0 to 0.65	1038@1			84.76% (10 000)@5	[120]
Bi_2S_3 nanorod/MoS_2 nanosheet	6 M KOH	−0.2 to 0.6	3040@1 and 1258@30			92.65% (5000) @10	[121]
MoS_2 nanosphere	6 M KOH	−0.2 to 0.6	1569@1 and 714@30			83.4% (5000)@10	
Bi_2S_3 nanorod	6 M KOH	−0.2 to 0.6	1007.8@1 and 389@30			74.5% (5000)@10	
ZnS/MnS	3 M KCl	− 0.8 to 0.8	884/2	7780	91	90% (5000)	[122]
CuS/MnS	3 M KOH	0 to 0.7	1144 @ 1 and 567 @ 50			85.9% (10 000)	[123]

mixed MS-based SCs. Among various MSs, nickel sulfide and cobalt sulfide have been widely explored for PSCs because of their easy synthesis methods, environmental friendliness, high electric conductivity, and the easy availability of their precursors. Nickel sulfide can be prepared in different forms, such as NiS, NiS_2, Ni_3S_2, Ni_3S_4, Ni_6S_5, Ni_7S_6, and Ni_9S_8, among which, NiS and Ni_3S_2 have been extensively reported for their use in SCs [90–92]. For example, Nandhinia *et al* prepared NiS, which showed a specific capacitance of 964 F g^{-1} at 1 A g^{-1} with a power density of 250 W kg^{-1}, an energy density of 16.5 Wh kg^{-1}, and a capacity retention of about 100% over 2000 cycles in 2 M KOH [90]. Zhang *et al* synthesized nanostructured Ni_3S_2 dendrites as an electrode for an SC, which exhibited a specific capacitance of 626.1 F g^{-1} at 5 A g^{-1} in 1 M KOH electrolyte [91]. Chen *et al* synthesized a self-supporting Ni_3S_2 nanosheet array with a superior specific capacitance of 1000 F g^{-1} at a high current density of 50 A g^{-1} in 1 M NaOH electrolyte [92]. Like nickel sulfide, cobalt sulfides also exist in a multitude of stoichiometries, such as CoS, CoS_2, Co_2S_3, Co_3S_4, Co_4S_3 and Co_9S_8, among which, CoS, CoS_2, and Co_9S_8, have been widely reported as electrode materials for SCs. For instance, Kumar *et al* prepared a hierarchical flower-like CoS nanostructure, which exhibited a specific capacitance of 570 F g^{-1} at 1 A g^{-1} with a capacity retention of about 97.9% over 2000 cycles in 6 M KOH electrolyte [93]. Ren *et al* reported hollow CoS_2 nanocubes with a specific capacitance of 936 F g^{-1} at 1 A g^{-1} in 2 M KOH electrolyte. The improved performance was attributed to the high surface area and an increased number of electrochemical active sites for the redox reaction [94]. Xing *et al* synthesized hierarchical mesoporous CoS_2 microspheres, which exhibited a specific capacitance of 718.7 F g^{-1} at 1 A g^{-1} in 2 M KOH electrolyte [95]. In addition to this, several studies of other metal sulfides, such as FeS_2, WS_2, MoS_2, SnS_2, VS_2, MnS, CuS, and Bi_2S_3 have also been published in the literature [96–104]. However, in their pure or intrinsic forms, the MSs used in electrodes only showed specific capacitances in the range of 200 to 1000 F g^{-1}. Also, most studies were carried out at low current densities and the cyclic stability was verified at about 3000 cycles. Several attempts have been made to obtain higher specific capacitance with more cyclic stability; the most successful were mixed or binary metal oxides and composites of an MS with another MS.

4.3.1 Mixed metal sulfides

Due to their low cost, structural scaffolds, narrow bandgap, high electrical conductivity, large anion polarizability, and the easy tailoring of their electrochemical properties, mixed MSs have emerged as promising candidates for SC applications. Mixed MSs also offer multiple oxidation states, which support reversible faradaic redox reactions, resulting in increased energy and power density [82, 85, 105, 106].

Research into several mixed MSs (binary and ternary) such as $NiCo_2S_4$, $CoNi_2S_4$, $ZnCo_2S_4$, $CuSbS_2$, $CoMoS_4$, $MnCo_2S_4$, $CuCo_2S_4$, NiV_2S_4, Cu_2NiSnS_4, and $FeNi_2S_4$ has shown that they have very high specific capacitance, electrical conductivity, energy density, power density, and cyclic stability. Similarly to the case of MSs, research into mixed MSs has also focused on synthesis by various methods and with different nanostructured morphologies in order to improve their electrochemical performance and process parameters [107–117].

For instance, Chen *et al* prepared ternary nickel cobalt sulfide interconnected nanosheet arrays on conductive carbon substrates by an electrodeposition method, which demonstrated a specific capacitance of 1418 F g^{-1} at 5 A g^{-1}, an energy density of 60 Wh kg^{-1} at a power density of 1.8 kW kg^{-1}, and a capacitance retention of 63.2% over 50 000 cycles. The mesoporous three-dimensional nano-architecture and network of the nanosheets were shown to be advantageous for such an enhancement in performance, as compared to nickel sulfide and cobalt sulfide [107]. Du *et al* reported a $CoNi_2S_4$ nanoparticle-based electrode for SCs with a specific capacitance of 1169 F g^{-1} at 1 A g^{-1} and an energy density of 46.9 Wh kg^{-1} at a power density of 446.6 W kg^{-1} in 3 M KOH electrolyte [108]. Hu *et al* prepared $CoNi_2S_4$ nanosheet arrays on nickel foams using a two-step hydrothermal method. An electrode made using the $CoNi_2S_4$ nanosheets showed a specific capacitance of 2906 F g^{-1} at a current density of 5 mA cm^{-2}, an energy density of 33.9 Wh kg^{-1} at a power density of 409 W kg^{-1}, and a retention of 78% over 3000 cycles in 2 M KOH electrolyte. This superior electrochemical performance was attributed to ravine-like nanosheet architectures with good mechanical and electrical contacts, which may have helped to enhance the charge transport rate and the ion diffusion process [109]. Zhu *et al* fabricated mesoporous $CuCo_2S_4$ nanoparticles by the solvothermal route, which exhibited a specific capacitance of 752 F g^{-1} at 2 A g^{-1} and a 98.1% retention of the initial specific capacitance after 5000 cycles in 2 M KOH electrolyte. The electrode also showed capacity retention of 47.9% over a wide range of current densities from 2 to 100 A g^{-1} [110].

Moosavifard *et al* successfully fabricated hierarchical hollow $CuCo_2S_4$ nano-needle arrays on Ni foam using a hydrothermal method, which exhibited a high specific capacitance of 2163 F g^{-1} at 60 A g^{-1}, an energy density of 44.1 Wh kg^{-1} at a power density of 800 W kg^{-1}, and a retention of 98.7% after 1000 cycles in 3 M KOH [111]. The electrolyte was also found to affect the electrochemical performance of $CuCo_2S_4$. $CuCo_2S_4$ with a polysulfide electrolyte (1 M KOH/1 M $Na_2S{\cdot}9H_2O$/ 1 M sulfur powder) showed an ultrahigh specific capacitance of 5030 F g^{-1} at a high current density of 20 A g^{-1} with a capacity retention of 79.5% after 2000 cycles [112]. Xu *et al* synthesized amorphous $CoMoS_4$ using a simple precipitation method, which exhibited a specific capacitance of 661 F g^{-1} at 1 A g^{-1}, an energy density of 27.2 Wh kg^{-1} at a power density of 400 W kg^{-1}, and a retention of 86% over 10 000 cycles [113]. Liu *et al* prepared flower-like manganese cobalt oxysulfide on a Ni foam using a two-step hydrothermal method, which exhibited specific capacitances of 490 C g^{-1} at 2 A g^{-1} and 415 C g^{-1} at 20 A g^{-1} with a capacity retention of 86.5% at a high current density of 20 A g^{-1} after 3000 cycles in 1 M KOH electrolyte. The flower-like structure provided a large number of surface active sites for electrochemical processes [114]. Sahoo and Rout fabricated ultrathin mesoporous manganese cobalt sulfide (MCS) nanosheet arrays on a Ni foam substrate by an electrodeposition method. An electrode made using the MCS nanosheets showed a specific capacitance of 2421 F g^{-1} at 1 A g^{-1} along with an energy density of 83.83 Wh kg^{-1} at a power of 5.65 kW kg^{-1} in 2 M KOH electrolyte. The authors argued that high porosity, a high surface area, and good electrical conductivity resulted in the high specific capacitance of the MCS nanosheet arrays [115]. Kuma *et al* synthesized NiV_2S_4 nanosheets on a 3D Ni framework, which exhibited a specific capacity of 639 C g^{-1} at 2 mA cm^{-2} in 6 M KOH electrolyte. The sample also showed an energy density of 45.1 Wh kg^{-1} at a power density of 240 W kg^{-1} and a retention of 90.7% after 2000 cycles [116].

Recently, Chandrasekaran *et al* studied a microwave-assisted $SnNi_2S_4$ (TNS) nanocomposite for SC applications. The hierarchical TNS nanocomposite showed a specific capacitance of 1484 F g^{-1} at 2 A g^{-1}, an energy density of 64.38 Wh kg^{-1} at 700.3 W kg^{-1}, and a potential window of 1.401 V in 1 M KOH electrolyte [117].

Several studies have reported composite metal sulfides with different MS combinations, which had enhanced performance compared to MSs in their intrinsic forms. A NiS/Ni_3S_4 composite showed much better performance than those of NiS or Ni_3S_4 alone [118]. The synergistic effect of the composite enabled better redox reactions and consequently enhanced the electrochemical performance of an SC.

For instance, Wang *et al* reported $Ni_3S_2@MoS_2$ core/shell nanorod arrays with more promising electrochemical performance reactions than those of Ni_3S_2, a single metal sulfide. An electrode made using the $Ni_3S_2@MoS_2$ core/shell nanorod arrays showed a specific capacitance of 848 F g^{-1}) at 5 A g^{-1}, which is about twice that of pristine Ni_3S_2 (425 F g^{-1}). Furthermore, at a higher current density of about 20 A g^{-1}, the sample exhibited an excellent rate capability of 46.6% and a capacitance retention of 91% after 2000 cycles. The improved performance of the core–shell structure was attributed to the active surface area available for ion diffusion due to the porous structure and the fast electron transfer enabled by the highly conductive nanorod core material [119].

Huang *et al* compared the electrochemical performances of MoS_2 and CoS_2 with that of a MoS_2-coated CoS_2 composite. The MoS_2-coated CoS_2 composite was prepared by a hydrothermal method. The estimated Brunauer–Emmett–Teller (BET) surface area of $CoS_2@MoS_2$ was found to be greater than those of MoS_2 and CoS_2 due to its nanoflower-like nanostructured morphology [120]. Figure 4.5 represents the electrochemical performance of MoS_2, CoS_2, and $CoS_2@MoS_2$ nanocomposite electrodes in a 2 M KOH electrolyte. CV curves at a scan rate of 5 mV s^{-1} in the potential range of 0.0 to 0.65 V and GCD plots at a current density of 1 A g^{-1} for different $CoS_2@MoS_2$ (CM1, CM2 and CM3) nanocomposites synthesized with different molar ratios of sodium molybdate dihydrate ($Na_2MoO_4.2H_2O$) and cobalt (II) nitrate hexahydrate ($Co(NO_3)_2.6H_2O$) are shown in figures 4.5(a) and (b). It can clearly be observed from figures 4.5(a) and (b) that the nanocomposite with a molar ratio of 1:2 has a much better performance than those of the other compositions. The electrochemical responses of MoS_2, CoS_2, and the $CoS_2@MoS_2$ nanocomposite (referred to as CM2 at a Mo:Co molar ratio of 2:1) was also compared. The CV curves in figure 4.5(c) for MoS_2, CoS_2, and the CM2 nanocomposite at a scan rate of 5 mV s^{-1} in a potential window of 0.0–0.65 V versus a saturated calomel electrode (SCE) show a larger integrated area for the nanocomposite (CM2), indicating that the nanocomposite has a higher charge storage capacity. The oxidation peak at around

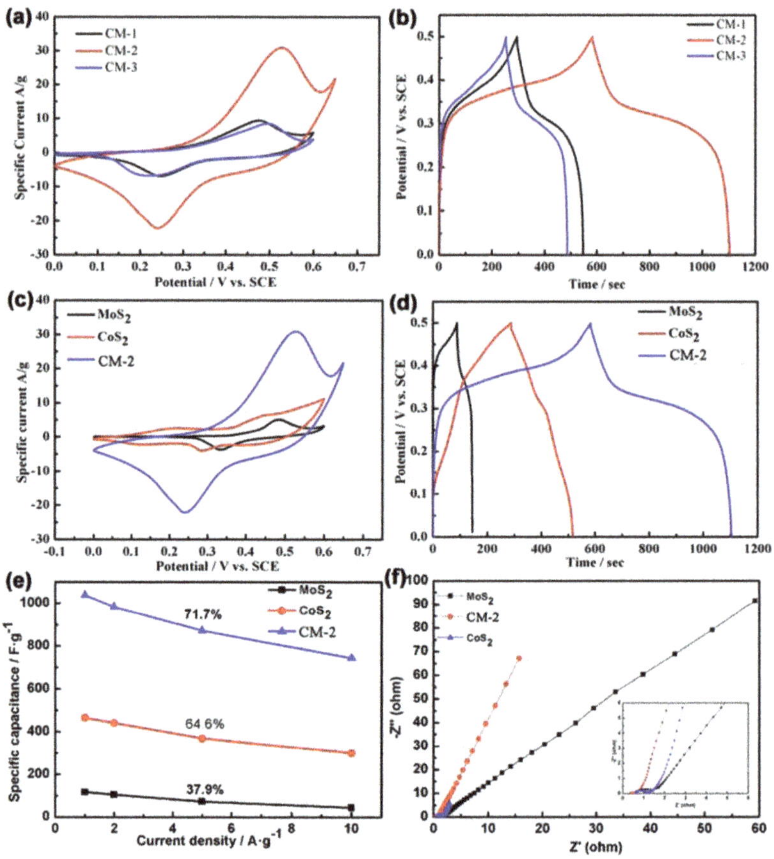

Figure 4.5. (a) CV curves at a scan rate of 5 mV s^{-1}, (b) GCD plots at a current density of 1 A g^{-1} for different $CoS_2@MoS_2$ (CM1, CM2 and CM3) nanocomposites, (c) CV curves at a scan rate of 5 mV s^{-1}, (d) GCD plots at a current density of 1 A g^{-1}, (e) rate capability curves at different current densities, and (f) EIS plots for MoS_2, CoS_2, and $CoS_2@MoS_2$. Reprinted from [120], Copyright (2018), with permission from Elsevier.

+0.52 V and the reduction peak at around +0.24 V for the $CoS_2@MoS_2$ nanocomposite are associated with faradaic redox reactions between the Co^{2+} and Co^{3+} states in KOH electrolyte. The authors proposed the following reaction mechanisms:

$$CoS_2 + OH^- \leftrightarrow CoS_2OH + e^- \tag{4.9}$$

$$MoS_2 + K^+ + e^- \leftrightarrow MoS^-SK^+. \tag{4.10}$$

The supercapacitive performance, as obtained from the GCD curves (figure 4.5(d)) of the MoS_2-coated CoS_2 composite, showed a specific capacitance of 1038 F g^{-1} at 1 A g^{-1}, which is much greater than those of MoS_2 (116.8) and CoS_2 (464.4 F g^{-1}). Furthermore, the $CoS_2@MoS_2$ nanocomposite electrode also showed an excellent rate capability, with a capacity retention of 71.7% over a wide range of current densities from 1 A g^{-1} to 10 A g^{-1}. The cyclic stability study shown in figure 4.5(e) uses GCD measurements between 0.0 and 0.5 V vs. SCE. It can be observed that the $CoS_2@MoS_2$ nanocomposite has a capacitance retention of 84.76% after 10 000 cycles at a current density of 5 A g^{-1}, which is much greater than those of MoS_2 and CoS_2, which have values of 57.02% and 80.71%, respectively. Also, in figure 4.5(f), which shows impedance plots, the measured equivalent series resistance (R_s) and charge transfer resistance (R_{ct}) of $CoS_2@MoS_2$ show very small values compared to those of MoS_2 and CoS_2. The authors argued that the improved performance of the $CoS_2@MoS_2$ nanocomposite could be attributed to:

- an improved ion diffusion rate due to the layered structure of MoS_2
- a larger electrochemical active surface area offered by the flower-like structure
- continuous channels for electron transportation offered by conductive CoS_2.

Fang *et al* prepared a 3D flower-like Bi_2S_3/MoS_2 nanohybrid by a facile one-pot hydrothermal method; the nanohybrid exhibited high specific capacitances of 3040 F g^{-1} at 1 A g^{-1} and 1258 F g^{-1} at 30 A g^{-1}, which are much greater than those of MoS_2 nanospheres (1569 F g^{-1}, 714 F g^{-1}, and 1007 F g^{-1}) and Bi_2S_3 nanorods (1007.8 F g^{-1} and 389 F g^{-1}). The Bi_2S_3/MoS_2 nanohybrid also showed a good capacitance retention of 92.65% after 5000 cycles at 10 A g^{-1} in 6 M KOH electrolyte [121]. Arul *et al* reported the electrochemical performances of ZnS, MnS, and ZnS/MnS nanocomposites prepared using a cost-effective chemical precipitation method. The ZnS/MnS nanocomposites exhibited a specific capacitance of 884 F g^{-1} at a scan rate of 2 mV s^{-1}, an energy density of 91 Wh kg^{-1} at a power density of 7.78 kW kg^{-1} in 3 M KCl electrolyte, and good cyclic stability with retention of 90% of the capacitance after 5000 cycles. According to the authors, the enhanced electrochemical performance was attributed to fast electron/ion transfer as a result of a synergetic effect in the wurtzite-type hexagonal structure of the ZnS/MnS nanocomposite [122]. Liu *et al* prepared a CuS/MnS composite with a stable three-dimensional clustered structure for SC applications. Their study revealed specific capacitances of 1144 F g^{-1} at 1 A g^{-1} and 567 F g^{-1} at 50 A g^{-1}, which are about five times higher than those of CuS or MnS. The composite sample also exhibited an excellent rate capability with a retention of 40.8% even at the very high current density of 100 A g^{-1} and good cyclic stability with a capacitive retention of 85.9% after 10 000 cycles [123]. In addition, several reports of composites of binary MSs with other binary or pure MSs are available in the literature; some of them are summarized in table 4.3.

Thus, in summary, studies of metal sulfides have mainly focused on enhanced redox chemistry, higher electrochemical activity (obtained by optimizing nanostructures), higher electronic conductivity, and low-cost material combinations. However, there is still some room for improvement.

4.4 Nanocomposite electrode materials for hybrid supercapacitors (HSCs)

In the previous sections, several composite electrodes (carbon–carbon, MO–MO, and MS–MS) have been discussed; the scope of those discussions was deliberately restricted to those particular classes only, in order to separate them from hybrid

Table 4.4. Electrochemical performance of HSCs based on MO@C and MS@C hybrid nanocomposite electrodes.

Working electrode	Electrolyte	Potential window (V)	Specific capacitance (F g^{-1}) at (@) current density (A g^{-1})/scan rate (mV s^{-1})	Power density (W kg^{-1})	Energy density (Wh kg^{-1})	Retention % (cycles) at (@) current density (A g^{-1})/scan rate (mV s^{-1})	References
Metal oxide@carbon nanocomposites							
Fe_3O_4@C	3.5 M KOH	−1.2 to 0.6	536@3 and 368@1	14 500	42	77% (20 000)@ 20	[140]
RuO_2@CNTs	0.5 M H_2SO_4	0 to 1	1170/10			82%/400	[141]
MnO_2@CNTs	1 M Na_2SO_4	0 to 0.9	642/10			100% (700)	[142]
MnO_2@CNTs	1 M Na_2SO_4	−0.2 to 0.8	201@1	600	13.3	100% (10 000)@1	[142]
Amorphous MnO_2@GO	1 M Li_2SO_4	0 to 0.8	372@0.5			92% (4000)@0.5	[143]
MnO_2@porous grapheme	1 M H_2SO_4	0 to 2	256@0.25	24 500	20.8	87.7% (1000)	[144]
Mn_3O_4@graphene sheets	6 M KOH	−0.8 to 0.2	271.5@0.1			100% (20 000)	[145]
RuO_2@graphene sheets	1 M H_2SO_4	−0.2 to 0.8	570/1	10 000	20.1	97.9% (1000)@1	[146]
Co_3O_4 nanoplates@graphene sheets	2 M KOH	−0.1 to 0.5	667.9@1.25			81.3% (1000)	[147]
ZnO@graphene	1 M KOH	−0.6 to 0.3	84/10			87.5% (1000)@2	[147]
MnO_2@rGO hydrogel and aerogel	1 M Na_2SO_4	0 to 1.6	242@1	820	21.2	89.6% (1000)	[149]
Single-crystal Fe_2O_4@ graphene	1 M KOH	−1.2 to −0.5	908@2			69%@50	[150]
MnO_2@GPCNs	1 M Na_2SO_4	0 to 1	438@0.5	47 500	50.2	91.1% (1000)	[151]
Mn_3O_4@CNFs	1 M Na_2SO_4	0 to 1	300.7/5		0.0794	100% (7500)	[152]
RuO_2@CNFs	6 M KOH	−0.2 to 0.8	188@at 1 mA cm^{-2}			93% (3000)	[153]
Fe_2O_3@graphene @CNTs	6 M KOH	0 to 0.4	675.7@1		60.3	82.2% (5000)@5	[154]
WS_2/$NiCo_2O_4$@C	3 M KOH	−0.4 to 0.6	2.449 F $cm^{-2@}$1 mA cm^{-2}	992.83	45.67	85.59% (5000)@0.02	[155]
VO_2/graphene@NiS_2	PVA/KOH gel	0 to 0.6	1280 @1	350	60.2	86.2% (10 000)	[156]
Zn_2SnO_4/SnO2@CNTs	6 M KOH	0 to 1	702@1	1000	98	93% (15 000)	[157]

(*Continued*)

Table 4.4. (*Continued*)

Working electrode	Electrolyte	Potential window (V)	Specific capacitance (F g^{-1}) at (@) current density (A g^{-1})/scan rate (mV s^{-1})	Power density (W kg^{-1})	Energy density (Wh kg^{-1})	Retention % (cycles) at (@) current density (A g^{-1})/scan rate (mV s^{-1})	References
Metal sulfide@carbon nanocomposites							
C@ $NiCo_2S_4$ nanorods	6 M KOH	0 to 0.5	1182@1			83% (2000)	[158]
NiS@C microspheres	6 M KOH	0 to 0.6	1594@ 1	800	19.5		[159]
MoS_2@AC	1 M Na_2SO_4	−0.8 to 0.8	261/2				[160]
MoS_2/MoOx@AC	1 M H_2SO_4	−0.8 to 0.8	230/5			128% (1500)	[161]
MoS_2@AC (corncob-derived)	1 M Na_2SO_4	−0.4 to 0.5	333.5@1	608	7.6	72.7% (10 000)	[162]
NiS_2/CoS_2@N-doped carbon	6 M KOH	0 to 0.6	1325@1	800	53.93	75% (5000)	[163]
CoS_2@S, N co-doped porous carbon	2 M KOH	−0.2 to 0.4	1148@2	725	56.4	94.1% (5000)	[164]
MoS_2/FeS_2@carbon nanospheres	6 M KOH	0 to 0.4	1321@2			81.2% (1000)	[165]
NiS@CNTs	polysulfide	−0.2 to 0.6	398.16@1 mA cm^{-2}	145.45	35.39	98.39% (1000)	[166]
CoS@CNTs	polysulfide	−0.2 to 0.6	288.43@1 mA cm^{-2}	145.45	25.63	95.3% (1000)	
CuS@CNTs	polysulfide	−0.2 to 0.6	176.20@1 mA cm^{-2}	145.45	15.66	78.57% (1000)	
PbS@CNTs	polysulfide	−0.2 to 0.6	101.94@1 mA cm^{-2}	145.45	9.06	69.49% (1000)	
NiS /MoS_2 @CNTs	3 M KOH	0 to 0.6	108@0.5		40	100% (10 000)	[167]
MnS_2@CNTs	1 M Na_2SO_4	−0.8 to 0.4	855@1 mA cm^{-2}			90.1% (5000)	[168]
MnS_2@CNTs	1 M KOH	−0.8 to 0.4	669@1 mA cm^{-2}			86.4% (5000)	
$NiCo_2S_4$@ CNTs	2 M KOH	0 to 0.5	1206@1		45.35	91.33% (8000)	[169]
ZnS@graphene	6 M KOH	−0.2 to 0.2	197.1/5			94.1% (1000)	[170]
$NiCo_2S_4$@graphene	6 M KOH	−0.055 to 0.55	1021@20	177.8	39.5	84.4% (15 000)	[171]
$NiCo_2S_4$@3Dgraphene (core/shell structure)	3 M KOH	0 to 1.6	1950@0.01	2460	23.9	93% (5000)	[172]
$CuCo_2S_4$@graphene	3 M KOH	0.01 to 0.6	665 //7.5			83% (1000)	[173]
MoS_2@graphene	1 M Na2SO4	0 to −1.0	401.1@1	1015.4	26.4	95% (1000)	[174]
MoS_2@graphene	1 M KOH	−0.25 to −0.75	756@0.5	125	26.6	88% (10 000)	[175]

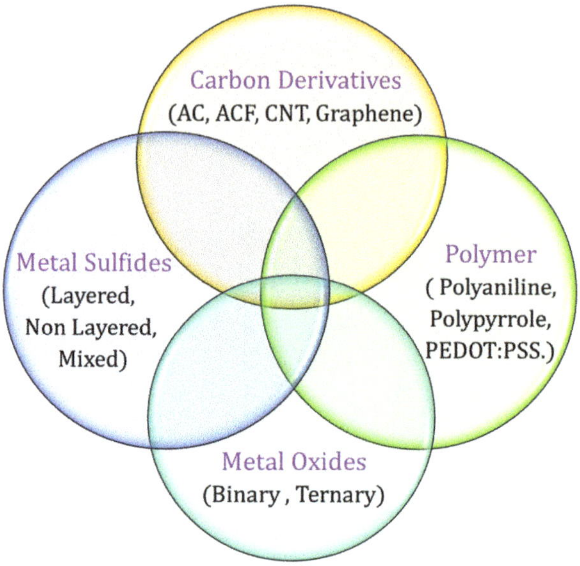

Figure 4.6. Schematic of the possible compositions of hybrid electrode materials used for HSCs.

electrodes for HSCs. Based on the type of HSC, the possible and explored combinations of hybrid electrode materials are shown in figure 4.6.

Carbon in different forms, such as activated carbon, graphene, carbon fiber, carbon nanotubes, and their composites with each other have been used as pioneer electrode materials for EDLCs. Several EDLCs that use activated carbon and graphene have been commercialized. As the name suggests, in EDLCs, charges are stored by the formation of an electrochemical double layer, however, the presence of functional groups on the surface also makes a pseudocapacitive contribution to the total specific capacitance. The electrochemical processes are mainly decided by the accessibility of the electrochemical active sites in the electrode, which are further decided by its porous network. Carbon materials are beneficial to the easy tailoring of porous networks. Also, the mechanical strength of carbon electrodes has been found to be greater as compared to other pseudocapacitive-material-based electrodes. Thus, proposals have been made to enhance the electrochemical properties of SCs using composites of carbon with other materials.

Several experimental and theoretical studies have recently been carried out based on this proposition, and have been found to work well, resulting in giant electrochemical performance in HSCs. These continuing trials will hopefully arrive at a synchronized outcome, which will overcome the drawbacks of the present commercial forms of SCs and batteries [124–128].

4.4.1 Metal oxide–carbon composite and metal sulfide–carbon composite electrode materials for HSCs

In the case of metal oxides, the charge transfer resistance can be decreased by increasing the active surface. However, because of their poor structural stability and mechanical strength, they are not able to provide good cyclic stability. Composites of MOs with carbon (MO@C) not only have increased mechanical strength but also enhanced electrical conductivity, and hence, supercapacitance. In MO@C, metal oxide nanostructures are developed on a carbon template. The carbon material acts as a porous network of electrochemical active sites, which helps to ease the process of charge and ion transportation. The carbon material also enhances the mechanical strength of the electrode. Many carbon derivatives with various nanostructured MOs have been reported recently as hybrid electrodes for HSCs.

Let us consider a typical example of RuO_2, one of the most efficient materials for PSCs. In spite of its superior capabilities, its cost and the low specific capacitance of the highly crystalline form of RuO_2 limit its large-scale application. Hybrid electrodes made from RuO_2 and carbonaceous materials reduce the material costs without suppressing the electrochemical activity [129].

A RuO_2–graphene nanosheet–CNT composite electrode was fabricated using graphene nanosheets (GNSs) and CNTs, which were deposited onto CNFs. The combination of RuO_2 and the carbon-based material provided a specific capacitance of 480.3 F g^{-1} at 0.6 A g^{-1}, which was much higher than those of RuO_2, GNSs, and CNTs alone. Furthermore, a RuO_2@GC-based SC exhibited a high energy density of 30.9 Wh kg^{-1} at a power density of 14 000 W kg^{-1} [130]. However, an increased specific surface area or pore volume does not ensure a high specific capacitance. It has been noted that the electrochemical performance of a Ru-based composite is strongly influenced by the percentage of ruthenium, as the active surface area and accessible pores are found to be strongly influenced by the amount of Ru.

CNTs are another favorable option for hybrid composite electrodes because of their low mass density, high conductivity, and hierarchical tubular channeled structure with an increased surface area for anchoring other materials onto them. Aligned CNTs with hollow-shelled RuO_2 exhibited a capacitance of 819 F g^{-1} at 0.2 A g^{-1}. The higher specific capacitance was attributed to the synergic effect of CNT and RuO_2 due to the dispersion of RuO_2 on the CNT network [131].

Likewise, MnO_2 in its pristine form does not have structural stability over a number of charge–discharge cycles. Compositing MnO_2 with CNTs helped to maintain its structural stability and good capacitance retention over a large number of cycles [132]. An asymmetrical SC made using a MnO_2@CNT composite exhibited a specific capacitance of 201 F g^{-1} at 1 A g^{-1} with an energy density of 13.3 Wh kg^{-1} at a power density of 600 W kg^{-1} [133]. In the nanocomposite, the CNTs not only increased the active sites available for faradaic processes but also provided a conductive network for a significant amount of charge transport, which resulted in a higher specific capacitance. The enhanced performance was also attributed to improved conductivity and a better ion diffusion process [134].

The problem of poor electron transport in Fe_2O_3 was addressed by the incorporation of MWCNTs into α-Fe_2O_3 nanospheres. A hybrid electrode made using the Fe_2O_3@MWCNTs nanocomposite showed an energy density of 50 Wh kg^{-1} at a power density of 1000 W kg^{-1} over a wide potential range of 0 to 2.8 V [135].

Graphene has also been reported to be a promising candidate for composites, due to its chemical stability and low cost. Irrespective of its good mechanical flexibility, high specific surface area of about ~2630 $m^2\ g^{-1}$, and good electrical conductivity, graphene in its pure form does not exhibit good electrochemical performance because of its agglomeration and sheet restacking problems. Composites of MOs or MSs with graphene have been reported to have superior capacitance, even with reduced electrode thicknesses, because of the layered structure of graphene. For instance, a MnO_2@graphene composite electrode exhibited a specific capacitance of ~216 F g^{-1}, which is much higher than those of either MnO_2 (~6.8 F g^{-1}) or graphene (~10.9 F g^{-1}) alone. An asymmetric SC with a MnO_2/graphene composite as the positive electrode and graphene as the negative electrode exhibited a superior energy density of ~42 Wh kg^{-1} and a power density of ~30 kW kg^{-1} [136].

Graphene-decorated V_2O_5 nanobelts were synthesized using the hydrothermal method, which revealed a high capacitance of 288 F g^{-1} at a scan rate of 10 mV s^{-1} and a retention of 82% after 5000 cycles [137]. Chen *et al* prepared a hybrid electrode material made from V_2O_5 nanoribbons and reduced graphene oxide (rGO) nanosheets using a simple cation-mediated self-assembly mechanism. The hybrid V_2O_5@rGO electrode showed a specific capacitance of 423.6 F g^{-1} at 1 A g^{-1}. A device developed using the hybrid material provided a high E_d of 58.8 Wh kg^{-1} and an average P_d of 500 W kg^{-1} as well as an excellent high rate capability. The specific capacitance remained nearly constant after up to 1800 cycles [138]. Likewise, graphene-decorated V_2O_5 nanobelt nanocomposite-based SCs exhibited a specific capacitance of 288 F g^{-1} at a scan rate of 10 mV s^{-1} with a capacitive retention of 82% over 5000 cycles [137].

Three-dimensional nanostructured carbon derivatives, such as carbon aerogels, mesoporous carbon, and carbon fiber have been used for composite electrodes in HSCs, as they offer a 3D porous network which results in exceptional electrochemical performance. For instance, Wang *et al* prepared a tungsten oxide@carbon aerogel (WO_3@C) nanocomposite dispersion of oxide particles in carbon aerogel, which exhibited good rate capacity and cyclic stability with a specific capacitance of 700 F g^{-1}. The specific capacitance of the WO_3@C aerogel nanocomposite was observed to be much higher than that of pristine tungsten oxide (54 F g^{-1}) [139].

Likewise, several other MO@C nanocomposites, such as Co_3O_4@C, IrO_2@C, CuO@C, ZnO@C, $NiCo_2O_4$@C, TiO_2@C, MoO_3@C, $VOPO_4$@c.c.@Fe_3O_4, RuO_2@CNT, MnO_2@CNT, MnO_2@GO, Mn_3O_4@graphene, RuO_2@graphene, Co_3O_4@graphene, ZnO@graphene, Fe_2O_4@graphene, Mn_3O_4@CNFs, RuO_2@CNFs, Zn_2SnO_4@SnO_2-CNTs, etc, have been reported to create a synergic improvement in electrochemical performance. Table 4.4 summarizes the electrochemical performance of hybrid nanocomposite electrodes fabricated with different MOs and MSs and porous carbon derivatives (AC, CNTs, graphene, rGO, CNFs, etc).

In addition to MO@C, nanocomposites of some transition-metal sulfides (MS@C) with AC, CNTs, graphene, rGO, and CNFs have also been observed to be attractive candidates for use in HSCs. Most of them have high abundance, good electrical conductivity, and strong interfacial interactions, as compared to metal oxides. The incorporation of nanostructured transition-metal sulfides into a porous carbon matrix has recently offered more promising electrochemical properties than those of MO@C composites. The significantly improved performance of HSCs with MS@C is attributed to the synergic effect of MSs and carbon materials. For example, carbon or its derivatives endow MS@C with more ion or electron paths and excellent chemical as well as mechanical stability. Carbon also protects the nanocomposite from corrosion by the electrolyte. The MS structure provides an increased number of electrochemical active sites and a hierarchical porous network for fast kinetic processes. It also hinders the volume change during charge–discharge cycling, and hence enhances the lifetime of HSCs. Several MS@C nanocomposites, such as $NiCo_2S_4$@C, NiS@C, MoS_2@C, CoS_2@C, FeS_2@C, NiS@CNT, CoS@CNT, CuS@CNT, PbS@CNT, MnS_2@CNT, ZnS@graphene, $CuCo_2S_4$@graphene, MoS_2@graphene, etc, have been explored for use as electrodes in all classes of HSC [158–175]. The electrochemical performances of some of them are summarized in table 4.4.

In summary, for HSCs, the selection of materials for composite electrodes and the enhancement of electrochemical performance depend on the following properties:

1. Structural, mechanical, thermal, and chemical stability
2. Electrical conductivity
3. Rapid ionic diffusion as well as ion transport through a network of inter- and intraplanar pores
4. Existence of (multiple) oxidation state(s)
5. High surface area and porosity
6. Pore volume, size, and distribution
7. Suitable surface wettability
8. Superior electrochemical activity
9. Synergic effect of composite elements for efficient charge transport.

The possible limitations of hybrid or composite materials include the following:

1. Nanostructured hybrid materials with organic and inorganic material composites are generally preferred. Therefore, more precautions or different methods are required during preparation processes in which conflicts may occur, such as thermal treatments: calcination, sintering, and annealing.
2. For bulk quantities, it is hard to maintain quality as well as the preferred composition in nanostructured hybrid materials
3. During scale-up, reproducibility and reparability problems may occur due to contamination and the accumulation of impurities.
4. The costs of materials and the production costs of complex composites may be problematic.

For commercialization, the key factors in reducing the price are:

1. The use of low-cost electrode materials, such as carbon black, and the use of low-cost, abundant, and environment-friendly precursors for carbon- and metal-based materials (either oxides or chalcogenides)
2. The design and development of automatic synthesis and assembly processes at a reasonable level of investment.

4.4.2 Battery-type Li-ion supercapacitors (LISCs)

The main motive for the development of electrochemical energy storage devices is to achieve high energy densities at high power densities and with longer cycle lives. Devices with such synchronized characteristics will be useful for future technological developments, such as hybrid electrical vehicles, the storage of renewable energy and electrical appliances in everyday life.

To date, SCs and lithium-ion batteries have been explored extensively to fulfill these needs. SCs storage charge using non-faradaic and faradaic pseudocapacitive processes. In EDLCs, charges are stored by electrostatic adsorption or desorption (anions and cations) onto the electrode surface. Due to the formation of the electric double layer, EDLCs have high power density with good cycle life, estimated to be about 10^5 cycles, but due to the limited accumulation of charge, they are unable to deliver high energy densities. In PSCs, charge is stored by a redox reaction between the electrode material and electrolyte ions. Pseudocapacitance can be produced by a redox reaction in the case of metal oxides, by a protonation reaction in the case of polymers, and by the reversible adsorption of hydrogen ions in the case of hydrated electrode materials.

However, LIBs store charge by the intercalation or de-intercalation of lithium ions in the host electrode materials, and thus, due to diffusion-controlled faradaic processes during cycling, LIBs can deliver high energy densities, but their performance is limited by low power density.

In the last two decades, massive efforts have been made to improve the performances of both SCs and LIBs, but still none of them has been found to be suitable for advanced future energy technologies. In order to synchronize power and energy densities, the hybridization of the working mechanisms of these two devices, i.e. SCs and LIBS, is opening a new era in electrochemical device development. Novel hybrid devices called lithium-ion SCs (LISCs) can be configured either in symmetric or asymmetric form, depending upon the combination of electrode materials used [176–180]. During charging, anions and cations move to the oppositely charged electrode, causing the accumulation of charges at the capacitive-type electrode and rapid redox reactions at the battery-type electrodes. As the electrodes work in different potential ranges, the working potential range is maximized.

In recent times, LISCs have emerged to bridge the gap between SCs and LIBs. The general configuration of LISCs is [176, 179]:

- a nanostructured anode with high electrical conductivity and high energy (e.g. a $Li_4Ti_5O_{12}$ (LTO) anode)
    - The high energy density of LISCs is based on Li^+ insertion in the anode, thus the chosen material should support the fast reversible intercalation of lithium ions
- a cathode with high power (e.g. an activated carbon (AC) cathode)
    - AC can not only serve as the cathode material but also as an anode material
- the electrolyte is an organic electrolyte with Li ions
    - non-aqueous electrolytes (usually $LiPF_6$ dissolved in ethylene carbonate and dimethyl carbonate).

To date, many research articles have been published on the subject of LISCs, which have mainly focused on novel anode and cathode materials. The ability of the surface of AC to adsorb electrostatic charge depends on its specific surface area and porosity. However, a focus on increasing this parameter of AC has been found to decrease its electrical conductivity and skeletal density. According to the literature, the poor rate capability and low cyclic stability of LISCs are mostly due to slow kinetics processes in the anode material. Previously, research into anode materials has focused on the development of hybrid electrode materials with different nanostructures and AC coatings. Hierarchical nanostructures reduce the charge diffusion path and enhance the active electrochemical sites, while a coating of conductive carbon offers a channeled framework for fast kinetic processes [176, 177, 181].

Similarly to the improvement of other hybrid capacitors, the specific capacitance of LISCs has been improved by modifying the surface structure and the porous network structure. The intercalation compounds for LISCs are:

1. Lithium-containing metal oxides
2. Polyanionic compounds
3. Graphene-based compounds
4. Binary and ternary metal oxides and their composite materials
5. Transition-metal sulfides and their nanocomposite materials.

Lithium-containing metal oxides can be used in two forms: layered ($LiMO_2$) oxides and spinel (LiM_2O_4) oxides (where M is a metal, such as Ni, Mn, Co, etc). Among various compounds, lithium cobalt oxide (LCO, $LiCoO_2$) is the most widely investigated, because of its high theoretical capacity of about 274 mAh g^{-1}. However, because of its structural instability during insertion and de-insertion, its actual performance is much lower than its theoretical predicted performance. Also, the toxic nature of cobalt has limited the further widespread application of $LiCoO_2$. Several efforts have been made to improve the performance of LCO, either by metal doping (doping with Zr, Mg, Mo, Sr, V, or Al) or surface modification using metal oxides such as TiO_2, Al_2O_3, SiO_2, B_2O_3, MgO, $LiMn_2O_4$, $AlPO_4$, and LiPON polyimide [180–187].

Studies of suitable combinations of Ni, Mn, and Co in $LiMO_2$ have described new compounds, such as $LiNi_{1/3}Co_{1/3}Mn_{1/3}O_2$ and $LiNi_{0.5}Mn_{1.5}O_4$, which have the potential to replace commercial $LiCoO_2$ [188–190]. For example, Brandt *et al* reported an LISC prepared using $LiNi_{0.5}Mn_{1.5}O_4$ and activated carbon with an energy densities of 40 Wh kg^{-1} and 63 Wh kg^{-1} at power densities of about 1 kW kg^{-1} and 100 W kg^{-1}, respectively. The device worked in the potential range from 0 to 3.3 V and exhibited a retention of 89%, even after 4000 cycles [189].

Another well studied class of materials used in LISCs is polyanionic compounds with the general form $Li_xA_y(BO_4)$, where A is a transition metal and B can be P, S, Si, etc. Because of their low cost, abundance, and environment- and eco-friendly features, several combinations of $Li_xA_y(BO_4)$ have been studied, among which, an olivine-type phosphate compound, $LiFePO_4$, has been thoroughly explored as a commercial LIB cathode because of its promising theoretical capacity of about

170 mAh g^{-1}. However, when compared to LCO, $LiFePO_4$ exhibited a lower electrochemical performance because of its low electric conductivity (about 10^{-9}–10^{-10} S cm^{-1}) and surface corrosion problem when it came into contact with the electrolyte [191–193]. Several efforts have been made to overcome these problems, either by the use of a conductive coating of carbon or by reduction of the particle size in order to improve the ionic conductivity and reduce surface corrosion [194, 195]. Satish *et al* synthesized carbon-coated $Li_3V_2(PO_4)_3$ (LVP-C) by the sol–gel technique, which showed an energy density of 27 Wh kg^{-1} at a power density of 255 kW kg^{-1} with a retention of 66% over 1000 cycles [195].

Metal oxides such as $Li_4Ti_5O_{12}$ have emerged as a new class of metal oxide electrode material for LISCs, known as zero-strain lithium insertion oxides, with a theoretical capacity of 175 mAh g^{-1} [194, 196–199]. Fan *et al* fabricated an LISC using an LTO nanosheet as the negative electrode and activated-nitrogen-doped graphene-based aerogel composites as the positive electrode, which exhibited an energy density of 70–21 Wh kg^{-1} at a power density of 200–8000 W kg^{-1} in a potential range of 1 to 3 V. The LISC also showed good cycling stability with a retention of 64% after 10 000 cycles [196]. However, their lower electrical conductivity hinders their future application, similarly to the case of $Li_xA_y(BO_4)$ compounds. Efforts have been made to improve the conductivity by reducing the particle size, making composites with other metal oxides, and coating the electrode with conductive carbon to improve its performance [200–202].

Nowadays, graphene is found to be a promising material for LISCs. Compared with other carbon derivatives, graphene has a higher electric conductivity, high electrochemical and thermal stability, flexibility, good mechanical strength, and a continuous porous network [203–206]. Also, graphene can be prepared in a variety of forms, such as reduced graphene oxide, graphene quantum dots, aerogel, hydrogel, 3D porous structures, etc. For instance, Kim *et al* prepared rGO-wrapped LTO, which delivered a maximum energy density of 50 Wh kg^{-1} at 16 W kg^{-1} in a voltage range of 1 to 2.5 V and retained up to 15 Wh kg^{-1}, even at a high power density of 2500 W kg^{-1}, with a retention of 75% over 1000 cycles [204]. Xu *et al* fabricated an LISC using $Li_4Ti_5O_{12}$–graphene (LTO@G) as the anode and AC as the cathode, which possessed a specific capacity of 194 mAh g^{-1} at 0.1C and exhibited energy densities of about 50–15 Wh kg^{-1} at power densities of 10–4000 W kg^{-1} in a potential range of 1 to 2.5 V [205]. Zhang *et al* reported a good rate capability for an LISC using an Fe_3O_4/graphene composite as the anode and porous graphene as the cathode. As a result, the prepared Fe_3O_4/graphene composite delivered a high capacity of more than 1000 mAh g^{-1} at 90 mA g^{-1} with energy densities of about 204 to 65 Wh kg^{-1} for power densities from 55 to 4600 W kg^{-1} and retained 90% of its charge after 1000 cycles [206].

Several other metal oxides, such as TiO_2, Fe_2O_3, MnO_2, V_2O_5, Bi_2O_3, Fe_3O_4, Nb_2O_5, $NiCo_2O_4$, $TiNb_2O_7$, $CdMn_2O_4$, $LiNbO_3$, Li_2TiO_3, $CoNiP_2O_7$, Li_2MnSiO_4, $Nb_{14}W_3O_{44}$, etc, and their composites with carbon and graphene materials have been explored for use in LISCs [207–220]. However, these components showed limited ionic diffusion in the active material and also possessed poor electrical conductivity that restricted their full utility for LISCs.

Recently, some transition-metal sulfides have attracted great attention in their binary and ternary forms as anodes for LISCs because of their good electrical conductivity and high theoretical predicted capacity values, as compared to the pioneer LTO [221–226]. Their structure also supports a rapid ion transfer process and a rapid intercalation mechanism. Sulphides such as MoS_2, CoS_2, $CoNi_2S_4$, TiS_2, etc, either in individual form with different structures, or in composite form with carbon materials, have been used as anodes for LISCs. For instance, an LISC with a nanocomposite of CNF@$CoNi_2S_4$ as the anode and AC as the cathode exhibited excellent energy densities of 85.4–35 Wh kg^{-1} at power densities of 150–15 000 W kg^{-1}. The device also showed excellent cyclic stability, with a retention of 96% after 5000 cycles at 2 A g^{-1}. The CNFs in the composite helped to improve the electrical conductivity and also reduced the stresses in the active electrode material that occur during the cycling process [224]. Wang *et al* synthesized a 3D hierarchical structure of CoS nanosheets by a hydrothermal method and used it as an anode material. An LISC fabricated with the CoS anode and a porous structured biochar of *fructus cannabis* shells (FCS) as the cathode showed a remarkable electrochemical performance with energy and power densities of 125.24 and 60 Wh kg^{-1} at power densities of 160 and 6400 W kg^{-1}, respectively, with a retention of 81.75% after 40 000 cycles. The enhanced performance was attributed to good kinetics due to the hierarchical structure of the CoS nanosheets and the synergy due to the matching of the hybrid's composition [226]. Some recent studies of LISCs using different electrode materials are summarized in table 4.5.

Thus, in summary, LISCs may become an ideal choice for applications such as electrical vehicles, for which both high energy and power density are required. LISC technology is still undergoing laboratory trials to optimize material performance and stability. The current research priority for LISC electrode materials should be the development of low-cost, abundant, and environment-friendly nanocomposite materials that exhibit high energy and power density with good electrochemical, thermal, structural, and long-term cyclic stability.

4.5 Summary

In this chapter, facts about the characteristics, merits, demerits, and electrochemical performance of pseudocapacitive and hybrid nanocomposite materials for SCs are discussed. Initially, research focused on enhancing the SSA and electrical conductivity of electrode materials. However, it can also be seen that the hierarchical surface morphology, pore size and shape, pore size distribution, and synchronization of the electrolyte ion and pore sizes also equally important to improving the performance of SCs. Three-dimensional surface morphological structures exhibit enhanced charge transfer and faster ion diffusion processes. All these things must be tailored for any class of electrode material intended for SC applications.

In the case of PSCs, the choice of material is based on three important parameters:

- electrical conductivity
- the oxidation states of the metal
- a structure that facilities intercalation into the oxide lattice

Table 4.5. Electrochemical performance of LISCs with different electrode materials.

LISC configuration						
Anode	Cathode	Potential window (V)	Maximum energy density, (Wh kg^{-1}) at [power density, (W kg^{-1})]	Energy density, (Wh kg^{-1}) at [maximum power density, (W kg^{-1})]	Cycle performance % retention (cycle)	References
$LiNi_{0.5}Mn_{1.5}O_4$	AC	0 to 3.3	63 [100]	40 [1000]	89% (4000)	[189]
$LiNi_{0.5}Mn_{1.5}O_4$	AC	1.5 to 3.25	19 [150]	8 [2500]	81% (3000)	[190]
$Li_3V_2(PO_4)_3$–C	AC	0.5 to 2.75	27 [300]		66% (1000)	[195]
LTO	N-Doped graphene	1 to 3	70 [200]	21 [8000]	64% (10 000)	[196]
LTO	AC derived from coconut shells	1 to 3	69 [500]		85% (2000)	[197]
LTO	rGO	1 to 3	45 [400]	30 [3300]	100% (5000)	[198]
LTO spheres	AC	1 to 3.5	74.3 [156.26]	41.7 [468.7]	93% (500)	[199]
Carbon@LTO	3D porous graphene	1.0 to 3.0	72 [650]	40 [8300]	65% (1000)	[200]
LTO@CNT	Graphene foam	1 to 3.6	101.8 [436.1]	12.7 [12 300]	84.8% (5000)	[201]
TiO_2@LTO	AC	0.5 to 2.5	74.85 [300]	36 [7500]	83.3% (5000)	[202]
Graphene@LTO	Graphene-sucrose	0 to 3	95 [45]	32 [3000]	94% (500)	[203]
Graphene-wrapped LTO	AC	1 to 2.5	50 [16]	15 [2500]	75% (1000)	[204]
Graphene@LTO	Graphene	0 to 3	95 [45]	32 [3000]	87% (500)	[203]
LTO@graphene	AC	1 to 2.5	50 [10]	15 [4000]	71% (10 000)	[205]
Fe_3O_4@graphene	3D graphene	1.0 to 4.0	147 [150]	86 [2587]	70% (1000)	[206]
TiO_2 nanobelt arrays	Graphene hydrogels	0 to 3.8	82 [570]	21 [19 000]	73% (600)	[207]
	AC	0 to 3	67.4 [75]	27.5 [5000]	80.5% (10 000)	[208]

(Continued)

Table 4.5. (*Continued*)

LISC configuration						
Anode	Cathode	Potential window (V)	Maximum energy density, (Wh kg^{-1}) at [power density, (W kg^{-1})]	Energy density, (Wh kg^{-1}) at [maximum power density, (W kg^{-1})]	Cycle performance % retention (cycle)	References
TiO_2@mesoporous carbon						
Graphene@TiO_2 hollow spheres	Graphene	0 to 3	72 [303]	10 [2000]	65% (1000)	[209]
TiO_2@rGO	AC	1 to 3	42 [800]	8.9 [8000]	80% (10 000)	[210]
F–Fe_2O_3	AC	0 to 3	28 [550]		90% (15 000)	[211]
T-Nb_2O_5@graphene	AC	0.8 to 3	47 [393]	15 [18 000]	93% (2000)	[212]
V_2O_5@CNT	AC	0 to 2.7	40 [210]	6.9 [6300]	78% (10 000)	[213]
MnO cubes	AC	0 to 4	227 [55]	21 [2952]	93% (3000)	[214]
Fe_xO@graphene	Graphene	0 to 3.5	129.6 [19]	45 [3500]	75% (3000)	[215]
$Li_3V_2(PO_4)_3$	AC	0 to 4	125 [300]	65 [6000]	80% (200)	[216]
$LiNbO_3$@graphene aerogel	Boron carbon nitride nanotubes	1 to 4	148 [200]	69.4 [9900]	82% (7000)	[217]
$NiCo_2O_4$	AC	0 to 4.5	39.4 [120]	10 [554]	100% (2000)	[218]
Li_2MnSiO_4	AC	0 to 3	54 [150]	37 [1500]	85% (1000)	[219]
$CoNiP_2O_7$	AC	0 to 4	116.3 [200]	66.7 [6486.5]	86.5% (500)	[220]
MoS_2 nanocrystal	AC	0.4 to 3.2		6 [5300]	90% (8000)	[221]
MoS_2@C-rGO	PANI-derived porous carbon	0 to 4	188 [200]	45.3 [40 000]	80% (10 000)	[222]
$CoMoS_4$	FCS	0 to 3	41.9 [75]	10 [3000]	95% (10 000)	[223]
CNF@$CoNi_2S_4$	AC	1 to 4	85.4 [150]	35 [15 000]	96% (5000)	[224]
TiS_2	AC	0 to 2.6	49 [100]		76% (2000)	[225]
3D hierarchical CoS	FCS	1 to 4.2	125.24 [160]	60 [6400]	81.75% (40 000)	[226]

The electrochemical performance, in terms of the specific capacitance, energy density, power density, rate capability, and cycling stability of MO- and MS-based SCs can be improved by optimizing the following factors:

- crystal structure
- percentage of crystallinity
- structural dimensions
- surface morphology
- chemical factors (the hydrous state of the oxide and the valence)
- electrical conductivity
- mass loading, i.e. the thickness of the electrode layer
- specific surface area
- pore size, shape, and pore size distribution
- electrolyte selection and its concentration.

Based on the synthesis conditions, the surface morphology and electrochemical behaviour can be drastically modified for each class of PSC material. Subsequently, the electrochemical performance of hybrid SCs with composite materials can be improved by optimizing the following key parameters:

- surface chemistry of the composite constituents
- nature of the chemical composition
- uniform and homogeneous pore size distribution
- surface morphology that provides an increased number of electrochemical active sites and channels for charge transport
- thermal, electrochemical, and mechanical stability
- a hierarchical 3D porous network (enhances active electrochemical sites and is also found to reduce the ion diffusion pathways)
- oriented nanostructures in the electrode material showing good conducting behaviour (and hence enhancing the rate capability and power density).

The carbon matrix in nanocomposites accommodates the structural expansion of MS or MO, helping to prevent the pulverization of nanoparticles and hence improving the specific capacitance and cyclic stability. In the case of MO@C and MS@C, well-distributed nanoparticles of metal oxides or metal sulfides on the carbon material's matrix surface create an enhanced surface area for electrochemical activity.

Lithium-metal-based materials are emerging as a promising class of electrode for hybrid SCs, which will boost the automotive electric vehicle industry. Researchers have mainly focused on the design and development of hybrid nanostructures with Li and suitable electrolytes for high energy, high power density, and longer cyclic performance of devices. The spaces between two adjacent particles in MO@C and MS@C nanocomposites offer space for the lithiation process in the case of LISCs, which improves its structural stability and hence cyclic stability.

Moreover, the future electrochemical performance of SC energy storage systems requires the consideration of the following aspects:

The chemical and physical interactions at the various interfaces of SCs have been shown to have a remarkable effect on their electrochemical performance. Hence, further fundamental study of the anisotropic growth of nanostructures and computational analysis of the different classes of components in SCs and their suitable combinations may yield more sustainable devices with enhanced electrochemical performance.

The unique designs and combinations of materials used in nanocomposites have shown significantly enhanced electrochemical properties and cyclic performance, both theoretically and at the lab scale, but they are still far from sufficient scalability and reproducibility.

Studies have shown that the synthesis route is critical when optimizing the crystallinity, morphology, porosity, dimensions, and hence the electrochemical properties of a material. Thus, a standard operational procedure for precise control of the synthesis parameters should be developed to support widespread industrial applications. Consequently, appropriate discrete tools for synthesis and characterization need to be developed. Also, a deep understanding of the factors influencing the electrode–electrolyte interaction needs to be explored further in order to identify suitable combinations.

As a consequence, considering such interesting features, the development of SCs will surely bring new dimensions to its next generation in the near future.

References and further reading

[1] Conway B E, Birss V and Wojtowicz J 1997 The role and utilization of pseudocapacitance for energy storage by supercapacitors *J. Power Sources* **66** 1–14

[2] González A, Goikolea E, Barrena J A and Mysyk R 2016 Review on supercapacitors: technologies and materials *Renew. Sustain. Energy Rev.* **58** 1189–206

[3] Wang Y, Song Y and Xia Y 2016 Electrochemical capacitors: mechanism, materials, systems, characterization and applications *Chem. Soc. Rev.* **45** 5925–50

[4] Liu J, Wang J, Xu C, Jiang H, Li C, Zhang L, Lin J and Shen Z X 2018 Advanced energy storage devices: basic principles, analytical methods, and rational materials design *Adv. Sci.* **5** 1700322

[5] Zhang Y, Feng H, Wu X, Wang L, Zhang A, Xia T, Dong H, Li X and Zhang L 2009 Progress of electrochemical capacitor electrode materials: a review *Int. J. Hydrogen Energy* **34** 4889–99

[6] Lee C Y and Bond A M 2010 Revelation of multiple underlying RuO_2 redox processes associated with pseudocapacitance and electrocatalysis *Langmuir* **26** 16155–62

[7] Adams D M, Brus L and Chidsey C E *et al* 2003 Charge transfer on the nanoscale: current status *J. Phys. Chem.* **B 107** 6668–97

[8] Krause P P, Camuka H, Leichtweiss T and Over H 2016 Temperature-induced transformation of electrochemically formed hydrous RuO_2 layers over Ru(0001) model electrodes *Nanoscale* **8** 13944–53

[9] Sugimoto W, Iwata H, Yokoshima K, Murakami Y and Takasu Y 2005 Proton and electron conductivity in hydrous ruthenium oxides evaluated by electrochemical impedance spectroscopy: the origin of large capacitance *J. Phys. Chem.* **B 109** 7330–8

[10] Gujar T P, Shinde V R, Lokhande C D, Kim W Y, Jung K D and Joo O S 2007 Spray deposited amorphous RuO_2 for an effective use in electrochemical supercapacitor *Electrochem. Commun.* **9** 504–10

[11] Kim H and Popov B N 2002 Characterization of hydrous ruthenium oxide/carbon nanocomposite supercapacitors prepared by a colloidal method *J. Power Sources* **104** 52–61

[12] Ramani M, Haran B S, White R E, Popov B N and Arsov L 2001 Studies on activated carbon capacitor materials loaded with different amounts of ruthenium oxide *J. Power Sources* **93** 209–14

[13] Cui G L, Zhou X H, Zhi L J, Thomas A and Müllen K 2007 Carbon/nanostructured Ru composites as electrodes for supercapacitors *New Carbon Mater.* **22** 302–6

[14] Liu Y, Zhao W and Zhang X 2008 Soft template synthesis of mesoporous Co_3O_4/ $RuO_2{\cdot}xH_2O$ composites for electrochemical capacitors *Electrochim. Acta* **53** 3296–304

[15] Lin K M, Chang K H, Hu C C and Li Y Y 2009 Mesoporous RuO_2 for the next generation supercapacitors with an ultrahigh power density *Electrochim. Acta* **54** 4574–81

[16] Zhang J, Ma J, Zhang L L, Guo P, Jiang J and Zhao X S 2010 Template synthesis of tubular ruthenium oxides for supercapacitor applications *J. Phys. Chem.* C **114** 13608–13

[17] Kuo S L and Wu N L 2003 Composite supercapacitor containing tin oxide and electroplated ruthenium oxide *Electrochem. Solid State Lett.* **6** A85

[18] Park B O, Lokhande C D, Park H S, Jung K D and Joo O S 2004 Performance of supercapacitor with electrodeposited ruthenium oxide film electrodes-effect of film thickness *J. Power Sources* **134** 148–52

[19] Wu X, Zeng Y, Gao H, Su J, Liu J and Zhu Z 2013 Template synthesis of hollow fusiform $RuO_2{\cdot}xH_2O$ nanostructure and its supercapacitor performance *J. Mater. Chem.* A **1** 469–72

[20] Wu X, Xiong W, Chen Y, Lan D, Pu X, Zeng Y, Gao H, Chen J, Tong H and Zhu Z 2015 High-rate supercapacitor utilizing hydrous ruthenium dioxide nanotubes *J. Power Sources* **294** 88–93

[21] Sugimoto W, Iwata H, Yasunaga Y, Murakami Y and Takasu Y 2003 Preparation of ruthenic acid nanosheets and utilization of its interlayer surface for electrochemical energy storage *Angew. Chem. Int. Ed.* **42** 4092–6

[22] Kim J Y, Kim K H, Kim H K, Park S H, Chung K Y and Kim K B 2014 Nanosheet-assembled 3D nanoflowers of ruthenium oxide with superior rate performance for supercapacitor applications *RSC Adv.* **4** 16115–20

[23] Sugimoto W, Shibutani T, Murakami Y and Takasu Y 2002 Charge storage capabilities of rutile-type RuO_2 VO_2 solid solution for electrochemical supercapacitors *Electrochem. Solid State Lett.* **5** A170

[24] Yokoshima K, Shibutani T, Hirota M, Sugimoto W, Murakami Y and Takasu Y 2006 Electrochemical supercapacitor behavior of nanoparticulate rutile-type $Ru_{1-x}V_xO_2$ *J. Power Sources* **160** 1480–6

[25] Yuan C Z, Gao B and Zhang X G 2007 Electrochemical capacitance of $NiO/Ru_{0.35}V_{0.65}O_2$ asymmetric electrochemical capacitor *J. Power Sources* **173** 606–12

[26] Takasu Y and Murakami Y 2000 Design of oxide electrodes with large surface area *Electrochim. Acta* **45** 4135–41

[27] Wu N L, Kuo S L and Lee M H 2002 Preparation and optimization of RuO_2-impregnated SnO_2 xerogel supercapacitor *J. Power Sources* **104** 62–5

[28] Hu C C, Yang Y L and Lee T C 2010 Microwave-assisted hydrothermal synthesis of $RuO_2{\cdot}xH_2O/TiO_2$ nanocomposites for high power supercapacitors *Electrochem. Solid State Lett.* **13** A173

[29] Jianming L and Xiaomei C 2015 RuO_2/MnO_2 composite materials for high-performance supercapacitor electrodes *J. Semiconduct.* **36** 083006
[30] Ji S H, Chodankar N R and Kim D H 2019 Aqueous asymmetric supercapacitor based on RuO_2-WO_3 electrodes *Electrochim. Acta* **325** 134879
[31] Li X, Zheng F, Luo Y, Wu Y and Lu F 2017 Preparation and electrochemical performance of TiO_2-SnO_2 doped RuO_2 composite electrode for supercapacitors *Electrochim. Acta* **237** 177–84
[32] Yong-gang W and Xiao-gang Z 2004 Preparation and electrochemical capacitance of RuO_2/TiO_2 nanotubes composites *Electrochim. Acta* **49** 1957–62
[33] Toupin M, Brousse T and Bélanger D 2004 Charge storage mechanism of MnO_2 electrode used in aqueous electrochemical capacitor *Chem. Mater.* **16** 3184–90
[34] Lindberg S, Jeschke S, Jankowski P, Abdelhamid M, Brousse T, Le Bideau J, Johansson P and Matic A 2020 Charge storage mechanism of α-MnO_2 in protic and aprotic ionic liquid electrolytes *J. Power Sources* **460** 228111
[35] Radhamani A V, Surendra M K and Rao M R 2018 Tailoring the supercapacitance of Mn_2O_3 nanofibers by nanocompositing with spinel-$ZnMn_2O_4$ *Mater. Des.* **139** 162–71
[36] Kang J, Chen L, Hou Y, Li C, Fujita T, Lang X, Hirata A and Chen M 2013 Electroplated thick manganese oxide films with ultrahigh capacitance *Adv. Energy Mater.* **3** 857–63
[37] Subramanian V, Zhu H and Wei B 2006 Nanostructured MnO_2: hydrothermal synthesis and electrochemical properties as a supercapacitor electrode material *J. Power Sources* **159** 361–4
[38] Wei J, Nagarajan N and Zhitomirsky I 2007 Manganese oxide films for electrochemical supercapacitors *J. Mater. Process. Technol.* **186** 356–61
[39] Teli A M, Beknalkar S A, Patil D S, Pawar S A, Dubal D P, Burute V Y, Dongale T D, Shin J C and Patil P S 2020 Effect of annealing temperature on charge storage kinetics of an electrospun deposited manganese oxide supercapacitor *Appl. Surf. Sci.* **511** 145466
[40] Singh A, Kumar D, Thakur A, Saini B S and Kaur R 2020 Enhanced performance of Fe-doped manganese oxide films as supercapacitor electrodes *Bull. Mater. Sci.* **43** 1–8
[41] Zhu S, Zhou H, Hibino M, Honma I and Ichihara M 2005 Synthesis of MnO_2 nano-particles confined in ordered mesoporous carbon using a sonochemical method *Adv. Funct. Mater.* **15** 381–6
[42] Ma S B, Nam K W, Yoon W S, Yang X Q, Ahn K Y, Oh K H and Kim K B 2007 A novel concept of hybrid capacitor based on manganese oxide materials *Electrochem. Commun.* **9** 2807–11
[43] Brousse T and Bélanger D 2003 A Hybrid Fe_3O_4 MnO_2 capacitor in mild aqueous electrolyte *Electrochem. Solid State Lett.* **6** A244
[44] Zarshad N, Wu J, Rahman A U and Ni H 2020 Fe-MnO_2 core–shell heterostructure for high-performance aqueous asymmetrical supercapacitor *J. Electroanal. Chem.* **871** 114266
[45] Arunpandiyan S, Vinoth S, Pandikumar A, Raja A and Arivarasan A 2021 Decoration of CeO_2 nanoparticles on hierarchically porous MnO_2 nanorods and enhancement of super-capacitor performance by redox additive electrolyte *J. Alloys Compd.* **861** 158456
[46] Zhao X, Hou Y, Wang Y, Yang L, Zhu L, Cao R and Sha Z 2017 Prepared MnO_2 with different crystal forms as electrode materials for supercapacitors: experimental research from hydrothermal crystallization process to electrochemical performances *RSC Adv.* **7** 40286–94
[47] Shin J, Seo J K, Yaylian R, Huang A and Meng Y S 2020 A review on mechanistic understanding of MnO_2 in aqueous electrolyte for electrical energy storage systems *Int. Mater. Rev.* **65** 356–87

[48] Kitchaev D A, Peng H, Liu Y, Sun J, Perdew J P and Ceder G 2016 Energetics of MnO_2 polymorphs in density functional theory *Phys. Rev.* B **93** 045132
[49] Zhang Y, Sun C, Lu P, Li K, Song S and Xue D 2012 Crystallization design of MnO_2 towards better supercapacitance *CrystEngComm* **14** 5892–97
[50] Wang H, Lu Z, Qian D, Li Y and Zhang W 2007 Single-crystal α-MnO_2 nanorods: synthesis and electrochemical properties *Nanotechnology* **18** 115616
[51] Park S, Shim H W, Lee C W, Song H J, Park I J, Kim J C, Hong K S and Kim D W 2015 Tailoring uniform γ-MnO_2 nanosheets on highly conductive three-dimensional current collectors for high-performance supercapacitor electrodes *Nano Res.* **8** 990–1004
[52] Wei C, Pang H, Zhang B, Lu Q, Liang S and Gao F 2013 Two-dimensional β-MnO_2 nanowire network with enhanced electrochemical capacitance *Sci. Rep.* **3** 1–5
[53] Zolfaghari A, Ataherian F, Ghaemi M and Gholami A 2007 Capacitive behavior of nanostructured MnO_2 prepared by sonochemistry method *Electrochim. Acta* **52** 2806–14
[54] Ma R E N Z H I, Bando Y, Zhang L I A N Q I and Sasaki T 2004 Layered MnO_2 nanobelts: hydrothermal synthesis and electrochemical measurements *Adv. Mater.* **16** 918–22
[55] Athouël L, Moser F, Dugas R, Crosnier O, Bélanger D and Brousse T 2008 Variation of the MnO_2 birnessite structure upon charge/discharge in an electrochemical supercapacitor electrode in aqueous Na_2SO_4 electrolyte *J. Phys. Chem.* C **112** 7270–7
[56] Ghodbane O, Pascal J L and Favier F 2009 Microstructural effects on charge-storage properties in MnO_2-based electrochemical supercapacitors *ACS Appl. Mater. Interfaces* **1** 1130–9
[57] Li F, Zhang Y X, Huang M, Xing Y and Zhang L L 2015 Rational design of porous MnO_2 tubular arrays via facile and templated method for high performance supercapacitors *Electrochim. Acta* **154** 329–37
[58] Wu M S 2005 Electrochemical capacitance from manganese oxide nanowire structure synthesized by cyclic voltammetric electrodeposition *Appl. Phys. Lett.* **87** 153102
[59] Hu C C and Tsou T W 2002 Ideal capacitive behavior of hydrous manganese oxide prepared by anodic deposition *Electrochem. Commun.* **4** 105–9
[60] Chen H, Qi X Q, Kuang M, Dong F and Zhang Y X 2016 Hierarchical copper/nickel-based manganese dioxide core–shell nanostructure for supercapacitor electrodes *Electrochim. Acta* **212** 671–7
[61] Xu J, Ding T, Wang J, Zhang J, Wang S, Chen C, Fang Y, Wu Z, Huo K and Dai J 2015 Tungsten oxide nanofibers self-assembled mesoscopic microspheres as high-performance electrodes for supercapacitor *Electrochim. Acta* **174** 728–34
[62] Tian J, Lin B, Sun Y, Zhang X and Yang H 2017 Porous WO_3@ CuO composites derived from polyoxometalates@ metal organic frameworks for supercapacitor *Mater. Lett.* **206** 91–4
[63] Chen D, Wang Q, Wang R and Shen G 2015 Ternary oxide nanostructured materials for supercapacitors: a review *J. Mater. Chem.* A **3** 10158–73
[64] Senthilkumar B, Sankar K V, Selvan R K, Danielle M and Manickam M 2013 Nano α-$NiMoO_4$ as a new electrode for electrochemical supercapacitors *RSC Adv.* **3** 352–7
[65] Yu X Y, Yao X Z, Luo T, Jia Y, Liu J H and Huang X J 2014 Facile synthesis of urchin-like $NiCo_2O_4$ hollow microspheres with enhanced electrochemical properties in energy and environmentally related applications *ACS Appl. Mater. Interfaces* **6** 3689–95
[66] Huang Z H, Sun F F, Batmunkh M, Li W H, Li H, Sun Y, Zhao Q, Liu X and Ma T Y 2019 Zinc–nickel–cobalt ternary hydroxide nanoarrays for high-performance supercapacitors *J. Mater. Chem.* A **7** 11826–35

[67] Khairy M and Mousa M 2019 Synthesis of ternary and quaternary metal oxides based on Ni, Mn, Cu, and Co for high-performance supercapacitor *J. Ovonic Res* **15** 181–98
[68] Biswas S, Sharma V, Mandal D, Chowdhury A, Chakravarty M, Priya S, Gowda C C, De P, Singh I and Chandra A 2020 Hollow nanostructures of metal oxides as emerging electrode materials for high performance supercapacitors *CrystEngComm* **22** 1633–44
[69] Li L, Peng S, Cheah Y, Teh P, Wang J, Wee G, Ko Y, Wong C and Srinivasan M 2013 Electrospun porous $NiCo_2O_4$ nanotubes as advanced electrodes for electrochemical capacitors *Chem.–A Eur. J.* **19** 5892–8
[70] An C, Wang Y, Huang Y, Xu Y, Xu C, Jiao L and Yuan H 2014 Novel three-dimensional $NiCo_2$ O_4 hierarchitectures: solvothermal synthesis and electrochemical properties *CrystEngComm* **16** 385–92
[71] Sahoo S, Naik K K and Rout C S 2015 Electrodeposition of spinel $MnCo_2O_4$ nanosheets for supercapacitor applications *Nanotechnology* **26** 455401
[72] Guo D, Luo Y, Yu X, Li Q and Wang T 2014 High performance $NiMoO_4$ nanowires supported on carbon cloth as advanced electrodes for symmetric supercapacitors *Nano Energy* **8** 174–82
[73] Mai L Q, Yang F, Zhao Y L, Xu X, Xu L and Luo Y Z 2011 Hierarchical $MnMoO_4$/$CoMoO_4$ heterostructured nanowires with enhanced supercapacitor performance *Nat. Commun.* **2** 1–5
[74] Lee H M, Gopi C V M, Rana P J S, Vinodh R, Kim S, Padma R and Kim H J 2018 Hierarchical nanostructured $MnCo_2O_4$–$NiCo_2O_4$ composites as innovative electrodes for supercapacitor applications *New J. Chem.* **42** 17190–94
[75] Gopi C V, Venkata-Haritha M, Kim S K, Prabakar K and Kim H J 2016 Flower-like ZnO@ $MnCo_2$ O_4 nanosheet structures on nickel foam as novel electrode material for high-performance supercapacitors *RSC Adv.* **6** 102961–7
[76] Wang L, Guan Y, Zhao X, Mu J, Che H, Li H and Guo Z 2018 $ZnCo_2O_4$@ $MnCo_2O_4$ heterojunction structured nanosheets for high-performance supercapacitor *J. Mater. Sci., Mater. Electron.* **29** 5782–90
[77] Zheng X, Ye Y, Yang Q, Geng B and Zhang X 2016 Hierarchical structures composed of $MnCo_2O_{4@}MnO_2$ core–shell nanowire arrays with enhanced supercapacitor properties *Dalton Trans.* **45** 572–8
[78] Feng X, Huang Y, Chen M, Chen X, Li C, Zhou S and Gao X 2018 Self-assembly of 3D hierarchical $MnMoO_4$/$NiWO_4$ microspheres for high-performance supercapacitor *J. Alloys Compd.* **763** 801–7
[79] Reddy A E, Anitha T, Gopi C V M, Durga I K and Kim H J 2018 Facile synthesis of hierarchical $ZnMn_2O_4$@$ZnFe_2O_4$ microspheres on nickel foam for high-performance supercapacitor applications *New J. Chem.* **42** 2964–9
[80] Gao X, Zhang Y, Huang M, Li F, Hua C, Yu L and Zheng H 2014 Facile synthesis of Co_3O_4@ $NiCo_2O_4$ core–shell arrays on Ni foam for advanced binder-free supercapacitor electrodes *Ceram. Int.* **40** 15641–6
[81] Geng P, Zheng S, Tang H, Zhu R, Zhang L, Cao S, Xue H and Pang H 2018 Transition metal sulfides based on graphene for electrochemical energy storage *Adv. Energy Mater.* **8** 1703259
[82] Yu X Y and Lou X W 2018 Mixed metal sulfides for electrochemical energy storage and conversion *Adv. Energy Mater.* **8** 1701592

[83] Theerthagiri J, Senthil R A, Nithyadharseni P, Lee S J, Durai G, Kuppusami P, Madhavan J and Choi M Y 2020 Recent progress and emerging challenges of transition metal sulfides based composite electrodes for electrochemical supercapacitive energy storage *Ceram. Int.* **46** 14317–45

[84] Jiang Y, Yang Y, Zhu J, Qiang L, Ye T, Li L, Su T and Fan R 2016 Nickel silicotungstate-decorated Pt photocathode as an efficient catalyst for triiodide reduction in dye-sensitized solar cells *Dalton Trans.* **45** 16859–68

[85] Sanchez J S, Pendashteh A, Palma J, Anderson M and Marcilla R 2019 Insights into charge storage and electroactivation of mixed metal sulfides in alkaline media: NiCoMn ternary metal sulfide nano-needles forming core–shell structures for hybrid energy storage *J. Mater. Chem.* A **7** 20414–24

[86] Jiang Z, Lu W, Li Z, Ho K H, Li X, Jiao X and Chen D 2014 Synthesis of amorphous cobalt sulfide polyhedral nanocages for high performance supercapacitors *J. Mater. Chem.* A **2** 8603–6

[87] Tao F, Zhao Y Q, Zhang G Q and Li H L 2007 Electrochemical characterization on cobalt sulfide for electrochemical supercapacitors *Electrochem. Commun.* **9** 1282–87

[88] Xu J, Wang Q, Wang X, Xiang Q, Liang B, Chen D and Shen G 2013 Flexible asymmetric supercapacitors based upon Co_9S_8 nanorod//Co_3O_4@ RuO_2 nanosheet arrays on carbon cloth *ACS Nano* **7** 5453–62

[89] Dou Y, Zhang L, Xu X, Sun Z, Liao T and Dou S X 2017 Atomically thin non-layered nanomaterials for energy storage and conversion *Chem. Soc. Rev.* **46** 7338–73

[90] Nandhini S and Muralidharan G 2018 Facile microwave-hydrothermal synthesis of NiS nanostructures for supercapacitor applications *Appl. Surf. Sci.* **449** 485–91

[91] Zhang Z, Huang Z, Ren L, Shen Y, Qi X and Zhong J 2014 One-pot synthesis of hierarchically nanostructured Ni_3S_2 dendrites as active materials for supercapacitors *Electrochim. Acta* **149** 316–23

[92] Chen J S, Guan C, Gui Y and Blackwood D J 2017 Rational design of self-supported Ni_3S_2 nanosheets array for advanced asymmetric supercapacitor with a superior energy density *ACS Appl. Mater. Interfaces* **9** 496–504

[93] Kumar K A, Pandurangan A, Arumugam S and Sathiskumar M 2019 Effect of bi-functional hierarchical flower-like CoS nanostructure on its interfacial charge transport kinetics, magnetic and electrochemical behaviors for supercapacitor and DSSC applications *Sci. Rep.* **9** 1–16

[94] Ren J, Meng Q, Xu Z, Zhang X and Chen J 2019 CoS_2 hollow nanocubes derived from Co-Co Prussian blue analogue: high-performance electrode materials for supercapacitors *J. Electroanal. Chem.* **836** 30–7

[95] Xing J C, Zhu Y L, Li M Y and Jiao Q J 2014 Hierarchical mesoporous CoS_2 microspheres: morphology-controlled synthesis and their superior pseudocapacitive properties *Electrochim. Acta* **149** 285–92

[96] Chen J, Zhou X, Mei C, Xu J, Zhou S and Wong C P 2016 Pyrite FeS_2 nanobelts as high-performance anode material for aqueous pseudocapacitor *Electrochim. Acta* **222** 172–6

[97] Karade S S, Dwivedi P, Majumder S, Pandit B and Sankapal B R 2017 First report on a FeS-based 2 V operating flexible solid-state symmetric supercapacitor device *Sustain. Energy Fuels* **1** 1366–75

[98] Nagaraju C, Gopi C V M, Ahn J W and Kim H J 2018 Hydrothermal synthesis of MoS_2 and WS_2 nanoparticles for high-performance supercapacitor applications *New J. Chem.* **42** 12357–60

[99] Pandit B, Bommineedi L K and Sankapal B R 2019 Electrochemical engineering approach of high performance solid-state flexible supercapacitor device based on chemically synthesized VS_2 nanoregime structure *J. Energy Chem.* **31** 79–88
[100] Barik R, Devi N, Perla V K, Ghosh S K and Mallick K 2019 Stannous sulfide nanoparticles for supercapacitor application *Appl. Surf. Sci.* **472** 112–17
[101] Miniach E and Gryglewicz G 2018 Solvent-controlled morphology of bismuth sulfide for supercapacitor applications *J. Mater. Sci.* **53** 16511–23
[102] Li D, Song S, Lu J, Liang J, Zhang Y and Li L 2019 A general self-template-etched solution route for the synthesis of 2D γ-manganese sulfide nanoplates and their enhanced supercapacitive performance *New J. Chem.* **43** 4674–80
[103] Heydari H, Moosavifard S E, Elyasi S and Shahraki M 2017 Nanoporous CuS nano-hollow spheres as advanced material for high-performance supercapacitors *Appl. Surf. Sci.* **394** 425–30
[104] Zhang Y, Xu J, Zheng Y, Hu X, Shang Y and Zhang Y 2016 Interconnected CuS nanowalls with rough surfaces grown on nickel foam as high-performance electrodes for supercapacitors *RSC Adv.* **6** 59976–83
[105] Nguyen T T, Balamurugan J, Aravindan V, Kim N H and Lee J H 2019 Boosting the energy density of flexible solid-state supercapacitors via both ternary NiV_2Se_4 and $NiFe_2Se_4$ nanosheet arrays *Chem. Mater.* **31** 4490–504
[106] Balamurugan J, Li C, Aravindan V, Kim N H and Lee J H 2018 Hierarchical Ni-Mo-S and Ni- Fe-S nanosheets with ultrahigh energy density for flexible all solid-state supercapacitors *Adv. Funct. Mater.* **28** 1803287
[107] Chen W, Xia C and Alshareef H N 2014 One-step electrodeposited nickel cobalt sulfide nanosheet arrays for high-performance asymmetric supercapacitors *ACS Nano* **8** 9531–41
[108] Du W, Zhu Z, Wang Y, Liu J, Yang W, Qian X and Pang H 2014 One-step synthesis of $CoNi_2S_4$ nanoparticles for supercapacitor electrodes *RSC Adv.* **4** 6998–7002
[109] Hu W, Chen R, Xie W, Zou L, Qin N and Bao D 2014 $CoNi_2S_4$ nanosheet arrays supported on nickel foams with ultrahigh capacitance for aqueous asymmetric supercapacitor applications *ACS Appl. Mater. Interfaces* **6** 19318–26
[110] Zhu Y, Ji X, Chen H, Xi L, Gong W and Liu Y 2016 The investigation of the electrochemically supercapacitive performances of mesoporous $CuCo_2S_4$ *RSC Adv.* **6** 84236–41
[111] Moosavifard S E, Fani S and Rahmanian M 2016 Hierarchical $CuCo_2S_4$ hollow nanoneedle arrays as novel binder-free electrodes for high-performance asymmetric supercapacitors *Chem. Commun.* **52** 4517–20
[112] Tang J, Ge Y, Shen J and Ye M 2016 Facile synthesis of $CuCo_2S_4$ as a novel electrode material for ultrahigh supercapacitor performance *Chem. Commun.* **52** 1509–12
[113] Xu X, Song Y, Xue R, Zhou J, Gao J and Xing F 2016 Amorphous $CoMoS_4$ for a valuable energy storage material candidate *Chem. Eng. J.* **301** 266–75
[114] Liu M, Fu Y, Ma H, Wang T, Guan C and Hu K 2016 Flower-like manganese-cobalt oxysulfide supported on Ni foam as a novel faradaic electrode with commendable performance *Electrochim. Acta* **191** 916–22
[115] Sahoo S and Rout C S 2016 Facile electrochemical synthesis of porous manganese-cobalt-sulfide based ternary transition metal sulfide nanosheets architectures for high performance energy storage applications *Electrochim. Acta* **220** 57–66

[116] Kumar R, Rai P and Sharma A 2016 Free-standing NiV_2S_4 nanosheet arrays on a 3D Ni framework via an anion exchange reaction as a novel electrode for asymmetric supercapacitor applications *J. Mater. Chem.* A **4** 17512–20

[117] Chandrasekaran N I, Muthukumar H, Sekar A D, Pugazhendhi A and Manickam M 2018 High-performance asymmetric supercapacitor from nanostructured tin nickel sulfide ($SnNi_2S_4$) synthesized via microwave-assisted technique *J. Mol. Liq.* **266** 649–57

[118] Gou J, Xie S, Yang Z, Liu Y, Chen Y, Liu Y and Liu C 2017 A high-performance supercapacitor electrode material based on NiS/Ni_3S_4 composite *Electrochim. Acta* **229** 299–305

[119] Wang J, Chao D, Liu J, Li L, Lai L, Lin J and Shen Z 2014 Ni_3S_2@ MoS_2 core/shell nanorod arrays on Ni foam for high-performance electrochemical energy storage *Nano Energy* **7** 151–60

[120] Huang F, Meng R, Sui Y, Wei F, Qi J, Meng Q and He Y 2018 One-step hydrothermal synthesis of a CoS_2@ MoS_2 nanocomposite for high-performance supercapacitors *J. Alloys Compd.* **742** 844–51

[121] Fang L, Qiu Y, Zhai T, Wang F, Lan M, Huang K and Jing Q 2017 Flower-like nanoarchitecture assembled from Bi_2S_3 nanorod/MoS_2 nanosheet heterostructures for high-performance supercapacitor electrodes *Colloids Surf.* A **535** 41–8

[122] Sabari Arul N, Cavalcante L S and Han J I 2018 Facile synthesis of ZnS/MnS nanocomposites for supercapacitor applications *J. Solid State Electrochem.* **22** 303–13

[123] Liu H, Guo Z, Wang X, Hao J and Lian J 2018 CuS/MnS composite hexagonal nanosheet clusters: synthesis and enhanced pseudocapacitive properties *Electrochim. Acta* **271** 425–32

[124] Haspert L C, Gillette E, Lee S B and Rubloff G W 2013 Perspective: hybrid systems combining electrostatic and electrochemical nanostructures for ultrahigh power energy storage *Energy Environ. Sci.* **6** 2578–90

[125] Liu Y, Hu P, Liu H, Song J, Umar A and Wu X 2019 Toward a high performance asymmetric hybrid capacitor by electrode optimization *Inorg. Chem. Front.* **6** 2824–31

[126] Liu H, Dai M, Zhao D, Wu X and Wang B 2020 Realizing superior electrochemical performance of asymmetric capacitors through tailoring electrode architectures *ACS Appl. Energy Mater.* **3** 7004–10

[127] Minakshi M, Mitchell D R, Jones R T, Pramanik N C, Jean-Fulcrand A and Garnweitner G 2020 A hybrid electrochemical energy storage device using sustainable electrode materials *ChemistrySelect* **5** 1597–606

[128] Choi H S and Park C R 2014 Theoretical guidelines to designing high performance energy storage device based on hybridization of lithium-ion battery and supercapacitor *J. Power Sources* **259** 1–14

[129] Hwang J Y, El-Kady M F, Wang Y, Wang L, Shao Y, Marsh K, Ko J M and Kaner R B 2015 Direct preparation and processing of graphene/RuO_2 nanocomposite electrodes for high-performance capacitive energy storage *Nano Energy* **18** 57–70

[130] Kong S, Cheng K, Ouyang T, Gao Y, Ye K, Wang G and Cao D 2017 Facile electrodepositing processed of RuO_2-graphene nanosheets-CNT composites as a binder-free electrode for electrochemical supercapacitors *Electrochim. Acta* **246** 433–42

[131] Wang P, Xu Y, Liu H, Chen Y, Yang J and Tan Q 2015 Carbon/carbon nanotube-supported RuO_2 nanoparticles with a hollow interior as excellent electrode materials for supercapacitors *Nano Energy* **15** 116–24

[132] Gao B, Li X, Ma Y, Cao Y, Hu Z, Zhang X, Fu J, Huo K and Chu P K 2015 MnO_2–TiO_2/C nanocomposite arrays for high-performance supercapacitor electrodes *Thin Solid Films* **584** 61–5

[133] Li L, Hu Z A, An N, Yang Y Y, Li Z M and Wu H Y 2014 Facile synthesis of MnO_2/CNTs composite for supercapacitor electrodes with long cycle stability *J. Phys. Chem.* C **118** 22865–72

[134] Lee J Y, Liang K, An K H and Lee Y H 2005 Nickel oxide/carbon nanotubes nanocomposite for electrochemical capacitance *Synth. Met.* **150** 153–7

[135] Zhao X, Johnston C and Grant P S 2009 A novel hybrid supercapacitor with a carbon nanotube cathode and an iron oxide/carbon nanotube composite anode *J. Mater. Chem.* **19** 8755–60

[136] Sun Y, Wu Q and Shi G 2011 Graphene based new energy materials *Energy Environ. Sci.* **4** 1113–32

[137] Lee M, Balasingam S K, Jeong H Y, Hong W G, Kim B H and Jun Y 2015 One-step hydrothermal synthesis of graphene decorated V_2O_5 nanobelts for enhanced electrochemical energy storage *Sci. Rep.* **5** 1–8

[138] Chen Y, Muthukumar K, Leban L II and Li J 2020 Microwave-assisted high-yield exfoliation of vanadium pentoxide nanoribbons for supercapacitor applications *Electrochim. Acta* **330** 135200

[139] Wang Y H, Wang C C, Cheng W Y and Lu S Y 2014 Dispersing WO_3 in carbon aerogel makes an outstanding supercapacitor electrode material *Carbon* **69** 287–93

[140] Kaipannan S, Govindarajan K, Sundaramoorthy S and Marappan S 2019 Waste toner-derived carbon/Fe_3O_4 nanocomposite for high-performance supercapacitor *ACS Omega* **4** 15798–805

[141] Kim I H, Kim J H and Kim K B 2005 Electrochemical characterization of electrochemically prepared ruthenium oxide/carbon nanotube electrode for supercapacitor application *Electrochem. Solid State Lett.* **8** A369

[142] Ren J, Li L, Chen C, Chen X, Cai Z, Qiu L, Wang Y, Zhu X and Peng H 2013 Twisting carbon nanotube fibers for both wire-shaped micro-supercapacitor and micro-battery *Adv. Mater.* **25** 1155–9

[143] Zhang B, Yu B, Zhou F and Liu W 2013 Polymer brush stabilized amorphous MnO_2 on graphene oxide sheets as novel electrode materials for high performance supercapacitors *J. Mater. Chem.* A **1** 8587–92

[144] Zhao X, Zhang L, Murali S, Stoller M D, Zhang Q, Zhu Y and Ruoff R S 2012 Incorporation of manganese dioxide within ultraporous activated graphene for high-performance electrochemical capacitors *ACS Nano* **6** 5404–12

[145] Zhu L, Zhang S, Cui Y, Song H and Chen X 2013 One step synthesis and capacitive performance of graphene nanosheets/Mn_3O_4 composite *Electrochim. Acta* **89** 18–23

[146] Wu Z S, Wang D W, Ren W, Zhao J, Zhou G, Li F and Cheng H M 2010 Anchoring hydrous RuO_2 on graphene sheets for high-performance electrochemical capacitors *Adv. Funct. Mater.* **20** 3595–602

[147] Wang X, Liu S, Wang H, Tu F, Fang D and Li Y 2012 Facile and green synthesis of Co_3O_4 nanoplates/graphene nanosheets composite for supercapacitor *J. Solid State Electrochem.* **16** 3593–602

[148] Wang J, Gao Z, Li Z, Wang B, Yan Y, Liu Q, Mann T, Zhang M and Jiang Z 2011 Green synthesis of graphene nanosheets/ZnO composites and electrochemical properties *J. Solid State Chem.* **184** 1421–7

[149] Wu S, Chen W and Yan L 2014 Fabrication of a 3D MnO_2/graphene hydrogel for high-performance asymmetric supercapacitors *J. Mater. Chem.* A **2** 2765–72

[150] Wang H, Xu Z, Yi H, Wei H, Guo Z and Wang X 2014 One-step preparation of single-crystalline Fe_2O_3 particles/graphene composite hydrogels as high performance anode materials for supercapacitors *Nano Energy* **7** 86–96

[151] Liu B, Liu Y, Chen H, Yang M and Li H 2019 MnO_2 nanostructures deposited on graphene-like porous carbon nanosheets for high-rate performance and high-energy density asymmetric supercapacitors *ACS Sustain. Chem. Eng.* **7** 3101–10

[152] Sambath Kumar K, Cherusseri J and Thomas J 2019 Two-dimensional Mn_3O_4 nanowalls grown on carbon fibers as electrodes for flexible supercapacitors *ACS Omega* **4** 4472–80

[153] Jeon S, Jeong J H, Yoo H, Yu H K, Kim B H and Kim M H 2020 RuO_2 nanorods on electrospun carbon nanofibers for supercapacitors *ACS Appl. Nano Mater.* **3** 3847–58

[154] Tian Y, Hu X, Wang Y, Li C and Wu X 2019 Fe_2O_3 nanoparticles decorated on graphene-carbon nanotubes conductive networks for boosting the energy density of all-solid-state asymmetric supercapacitor *ACS Sustain. Chem. Eng.* **7** 9211–19

[155] Li L, Gao J, Cecen V, Fan J, Shi P, Xu Q and Min Y 2020 Hierarchical WS_2@ $NiCo_2O_4$ core–shell heterostructure arrays supported on carbon cloth as high-performance electrodes for symmetric flexible supercapacitors *ACS Omega* **5** 4657–67

[156] Chen H C, Lin Y C, Chen Y L and Chen C J 2018 Facile fabrication of three-dimensional hierarchical nanoarchitectures of VO_2/graphene@NiS_2 hybrid aerogel for high-performance all-solid-state asymmetric supercapacitors with ultrahigh energy density *ACS Appl. Energy Mater.* **2** 459–67

[157] Samuel E, Kim T G, Park C W, Joshi B, Swihart M T and Yoon S S 2019 Supersonically sprayed Zn_2SnO_4/SnO_2/CNT nanocomposites for high-performance supercapacitor electrodes *ACS Sustain. Chem. Eng.* **7** 14031–40

[158] Li L, Dai Z, Zhang Y, Yang J, Huang W and Dong X 2015 Carbon@ $NiCo_2S_4$ nanorods: an excellent electrode material for supercapacitors *RSC Adv.* **5** 83408–14

[159] Zhang X, Lu Q, Liu H, Li K and Wei M 2020 Nature-inspired design of NiS/carbon microspheres for high-performance hybrid supercapacitors *Appl. Surf. Sci.* **528** 146976

[160] Sangeetha D N and Selvakumar M 2018 Active-defective activated carbon/MoS_2 composites for supercapacitor and hydrogen evolution reactions *Appl. Surf. Sci.* **453** 132–40

[161] Sari F N I and Ting J M 2018 MoS_2/MoOx-nanostructure-decorated activated carbon cloth for enhanced supercapacitor performance *ChemSusChem.* **11** 897–906

[162] Wang F, Ma J, Zhou K and Li X 2020 MoS_2/corncob-derived activated carbon for supercapacitor application *Mater. Chem. Phys.* **244** 122215

[163] Liu H, Guo H, Yue L, Wu N, Li Q, Yao W, Xue R, Wang M and Yang W 2019 Metal-organic frameworks-derived NiS_2/CoS_2/N-doped carbon composites as electrode materials for asymmetric supercapacitor *ChemElectroChem.* **6** 3764–73

[164] Liu S, Gao D, Li J, San Hui K, Yin Y, Hui K N and Jun S C 2019 Phosphorus dual-site driven CoS_2@S, N co-doped porous carbon nanosheets for flexible quasi-solid-state supercapacitors *J. Mater. Chem.* A **7** 26618–30

[165] Chen X, Shi T, Zhong K, Wu G and Lu Y 2020 Capacitive behavior of MoS_2 decorated with FeS_2@ carbon nanospheres *Chem. Eng. J.* **379** 122240

[166] Gopi C V M, Ravi S, Rao S S, Reddy A E and Kim H J 2017 Carbon nanotube/metal-sulfide composite flexible electrodes for high-performance quantum dot-sensitized solar cells and supercapacitors *Sci. Rep.* **7** 1–12

[167] Yang X, Zhao L and Lian J 2017 Arrays of hierarchical nickel sulfides/MoS_2 nanosheets supported on carbon nanotubes backbone as advanced anode materials for asymmetric supercapacitor *J. Power Sources* **343** 373–82

[168] Pujari R B, Lokhande A C, Shelke A R, Kale S B, Lee D W and Lokhande C D 2021 MnS_2/carbon nanotube electrode for improved supercapacitor performance *Solid State Sci.* **111** 106449

[169] Deng Y, Wang X, Wang Z, Wang X, Li Z, Wang L, Zhou C, Chen D and Luo Y 2021 Yolk–shell structured nickel cobalt sulfide and carbon nanotube composite for high-performance hybrid supercapacitors *Energy & Fuels* **35** 5342–51

[170] Ramachandran R, Saranya M, Kollu P, Raghupathy B P, Jeong S K and Grace A N 2015 Solvothermal synthesis of Zinc sulfide decorated Graphene (ZnS/G) nanocomposites for novel Supercapacitor electrodes *Electrochim. Acta* **178** 647–57

[171] Wang X, Zhao S X, Dong L, Lu Q L, Zhu J and Nan C W 2017 One-step synthesis of surface-enriched nickel cobalt sulfide nanoparticles on graphene for high-performance supercapacitors *Energy Storage Mater.* **6** 180–7

[172] Beka L G, Li X and Liu W 2017 Nickel cobalt sulfide core/shell structure on 3D Graphene for supercapacitor application *Sci. Rep.* **7** 1–11

[173] Annamalai K P and Tao Y S 2016 A hierarchically porous $CuCo_2S_4$/graphene composite as an electrode material for supercapacitors *New Carbon Mater.* **31** 336–42

[174] Fu M, Zhu Z, Chen W, Yu H and Liu Q 2020 Microwave-assisted synthesis of MoS_2/graphene composites for supercapacitors *J. Mater. Sci.* **55** 16385–93

[175] Vikraman D, Karuppasamy K, Hussain S, Kathalingam A, Sanmugam A, Jung J and Kim H S 2019 One-pot facile methodology to synthesize MoS_2-graphene hybrid nanocomposites for supercapacitors with improved electrochemical capacitance *Compos. B.* **161** 555–63

[176] He Z and Liu Y 2018 Fundamentals of lithium-ion supercapacitors *Lithium-Ion Supercapacitors* (Boca Raton, FL: CRC Press)

[177] Sun C, Zhang X, Li C, Wang K, Sun X and Ma Y 2020 High-efficiency sacrificial prelithiation of lithium-ion capacitors with superior energy-storage performance *Energy Storage Mater.* **24** 160–6

[178] Sui D, Wu M, Liu Y, Yang Y, Zhang H, Ma Y, Zhang L and Chen Y 2020 High performance Li-ion capacitor fabricated with dual graphene-based materials *Nanotechnology* **32** 015403

[179] Lin Y T, Chang-Jian C W, Hsieh T H, Huang J H, Weng H C, Hsiao Y S, Syu W L and Chen C P 2021 High-performance Li-ion capacitor constructed from biomass-derived porous carbon and high-rate $Li_4Ti_5O_{12}$ *Appl. Surf. Sci.* **543** 148717

[180] Needham S A, Wang G X, Liu H K, Drozd V A and Liu R S 2007 Synthesis and electrochemical performance of doped $LiCoO_2$ materials *J. Power Sources* **174** 828–31

[181] Furushima Y, Yanagisawa C, Nakagawa T, Aoki Y and Muraki N 2011 Thermal stability and kinetics of delithiated $LiCoO_2$ *J. Power Sources* **196** 2260–3

[182] Cho J and Kim G 1999 Enhancement of thermal stability of $LiCoO_2$ by $LiMn_2O_4$ coating *Electrochem. Solid State Lett.* **2** 253

[183] Valanarasu S and Chandramohan R 2010 Improvement of the cycle life of $LiCoO_2$ powder by Sr doping *J. Alloys Compd.* **494** 434–8

[184] Mladenov M, Stoyanova R, Zhecheva E and Vassilev S 2001 Effect of Mg doping and MgO-surface modification on the cycling stability of $LiCoO_2$ electrodes *Electrochem. Commun.* **3** 410–16

[185] Kim M G and Yo C H 1999 X-ray absorption spectroscopic study of chemically and electrochemically Li ion extracted $Li_yCo_{0.85}Al_{0.15}O_2$ compounds *J. Phys. Chem.* **B 103** 6457–65
[186] Kim Y J, Cho J, Kim T J and Park B 2003 Suppression of cobalt dissolution from the $LiCoO_2$ cathodes with various metal-oxide coatings *J. Electrochem. Soc.* **150** A1723
[187] Choi K H, Jeon J H, Park H K and Lee S M 2010 Electrochemical performance and thermal stability of $LiCoO_2$ cathodes surface-modified with a sputtered thin film of lithium phosphorus oxynitride *J. Power Sources* **195** 8317–21
[188] He Z, Wang Z, Chen H, Huang Z, Li X, Guo H and Wang R 2015 Electrochemical performance of zirconium doped lithium rich layered $Li_{1.2}Mn_{0.54}Ni_{0.13}Co_{0.13}O_2$ oxide with porous hollow structure *J. Power Sources* **299** 334–41
[189] Brandt A, Balducci A, Rodehorst U, Menne S, Winter M and Bhaskar A 2014 Investigations about the use and the degradation mechanism of $LiNi_{0.5}Mn_{1.5}O_4$ in a high power LIC *J. Electrochem. Soc.* **161** A1139
[190] Arun N, Jain A, Aravindan V, Jayaraman S, Ling W C, Srinivasan M P and Madhavi S 2015 Nanostructured spinel $LiNi_{0.5}Mn_{1.5}O_4$ as new insertion anode for advanced Li-ion capacitors with high power capability *Nano Energy* **12** 69–75
[191] Chang H H, Chang C C, Su C Y, Wu H C, Yang M H and Wu N L 2008 Effects of TiO_2 coating on high-temperature cycle performance of $LiFePO_4$-based lithium-ion batteries *J. Power Sources* **185** 466–72
[192] Cui Y, Zhao X and Guo R 2010 Enhanced electrochemical properties of $LiFePO_4$ cathode material by CuO and carbon co-coating *J. Alloys Compd.* **490** 236–40
[193] Guan X M, Li G J, Li C Y and Ren R M 2017 Synthesis of porous nano/micro structured $LiFePO_4$/C cathode materials for lithium-ion batteries by spray-drying method *Trans. Nonferrous Met. Soc. China* **27** 141–7
[194] Liu H, Liu X, Wang S, Liu H K and Li L 2020 Transition metal based battery-type electrodes in hybrid supercapacitors: a review *Energy Storage Mater.* **28** 122–45
[195] Satish R, Aravindan V, Ling W C and Madhavi S 2015 Carbon-coated $Li_3V_2(PO_4)_3$ as insertion type electrode for lithium-ion hybrid electrochemical capacitors: an evaluation of anode and cathodic performance *J. Power Sources* **281** 310–17
[196] Fan Q, Yang M, Meng Q, Cao B and Yu Y 2016 Activated-nitrogen-doped graphene-based aerogel composites as cathode materials for high energy density lithium-ion supercapacitor *J. Electrochem. Soc.* **163** A1736
[197] Jain A, Aravindan V, Jayaraman S, Kumar P S, Balasubramanian R, Ramakrishna S, Madhavi S and Srinivasan M P 2013 Activated carbons derived from coconut shells as high energy density cathode material for Li-ion capacitors *Sci. Rep.* **3** 1–6
[198] Aravindan V, Mhamane D, Ling W C, Ogale S and Madhavi S 2013 Nonaqueous lithium-ion capacitors with high energy densities using trigol-reduced graphene oxide nanosheets as cathode-active material *ChemSusChem.* **6** 2240–4
[199] Deng S, Li J, Sun S, Wang H, Liu J and Yan H 2014 Synthesis and electrochemical properties of $Li_4Ti_5O_{12}$ spheres and its application for hybrid supercapacitors *Electrochim. Acta* **146** 37–43
[200] Ye L, Liang Q, Lei Y, Yu X, Han C, Shen W, Huang Z H, Kang F and Yang Q H 2015 A high performance Li-ion capacitor constructed with $Li_4Ti_5O_{12}$/C hybrid and porous graphene macroform *J. Power Sources* **282** 174–8
[201] Liu Y, Wang W, Chen J, Li X, Cheng Q and Wang G 2020 Fabrication of porous lithium titanate self-supporting anode for high performance lithium-ion capacitor *J. Energy Chem.* **50** 344–50

[202] Gao L, Huang D, Shen Y and Wang M 2015 Rutile-TiO_2 decorated $Li_4Ti_5O_{12}$ nanosheet arrays with 3D interconnected architecture as anodes for high performance hybrid supercapacitors *J. Mater. Chem.* A **3** 23570–76

[203] Leng K, Zhang F, Zhang L, Zhang T, Wu Y, Lu Y, Huang Y and Chen Y 2013 Graphene-based Li-ion hybrid supercapacitors with ultrahigh performance *Nano Res.* **6** 581–92

[204] Kim H, Park K Y, Cho M Y, Kim M H, Hong J, Jung S K, Roh K C and Kang K 2014 High-performance hybrid supercapacitor based on graphene-wrapped $Li_4Ti_5O_{12}$ and activated carbon *ChemElectroChem.* **1** 125–30

[205] Xu N, Sun X, Zhang X, Wang K and Ma Y 2015 A two-step method for preparing $Li_4Ti_5O_{12}$–graphene as an anode material for lithium-ion hybrid capacitors *RSC Adv.* **5** 94361–8

[206] Zhang F, Zhang T, Yang X, Zhang L, Leng K, Huang Y and Chen Y 2013 A high-performance supercapacitor-battery hybrid energy storage device based on graphene-enhanced electrode materials with ultrahigh energy density *Energy Environ. Sci.* **6** 1623–32

[207] Wang H, Guan C, Wang X and Fan H J 2015 A high energy and power Li-ion capacitor based on a TiO_2 nanobelt array anode and a graphene hydrogel cathode *Small* **11** 1470–7

[208] Yang C, Lan J L, Liu W X, Liu Y, Yu Y H and Yang X P 2017 High-performance Li-ion capacitor based on an activated carbon cathode and well-dispersed ultrafine TiO_2 nanoparticles embedded in mesoporous carbon nanofibers anode *ACS Appl. Mater. Interfaces* **9** 18710–19

[209] Wang F, Wang C, Zhao Y, Liu Z, Chang Z, Fu L, Zhu Y, Wu Y and Zhao D 2016 A quasi-solid-state Li-ion capacitor based on porous TiO_2 hollow microspheres wrapped with graphene nanosheets *Small* **12** 6207–13

[210] Kim H, Cho M Y, Kim M H, Park K Y, Gwon H, Lee Y, Roh K C and Kang K 2013 A novel high-energy hybrid supercapacitor with an anatase TiO_2–reduced graphene oxide anode and an activated carbon cathode *Adv. Energy Mater.* **3** 1500–6

[211] Karthikeyan K, Amaresh S, Lee S N, Aravindan V and Lee Y S 2014 Fluorine-doped Fe_2O_3 as high energy density electroactive material for hybrid supercapacitor applications *Chem.–An Asian J.* **9** 852–7

[212] Kong L, Zhang C, Wang J, Qiao W, Ling L and Long D 2015 Free-standing T-Nb_2O_5/graphene composite papers with ultrahigh gravimetric/volumetric capacitance for Li-ion intercalation pseudocapacitor *ACS Nano* **9** 11200–8

[213] Chen Z, Augustyn V, Wen J, Zhang Y, Shen M, Dunn B and Lu Y 2011 High-performance supercapacitors based on intertwined CNT/V_2O_5 nanowire nanocomposites *Adv. Mater.* **23** 791–5

[214] Liu C, Zhang C, Song H, Zhang C, Liu Y, Nan X and Cao G 2016 Mesocrystal MnO cubes as anode for Li-ion capacitors *Nano Energy* **22** 290–300

[215] Li M, Pan F, Choo E S G, Lv Y, Chen Y and Xue J 2016 Designed construction of a graphene and iron oxide freestanding electrode with enhanced flexible energy-storage performance *ACS Appl. Mater. Interfaces* **8** 6972–81

[216] Böckenfeld N and Balducci A 2013 On the use of lithium vanadium phosphate in high power devices *J. Power Sources* **235** 265–73

[217] Jiang H, Wang S, Zhang B, Shao Y, Wu Y, Zhao H, Lei Y and Hao X 2020 High performance lithium-ion capacitors based on $LiNbO_3$-arched 3D graphene aerogel anode and BCNNT cathode with enhanced kinetics match *Chem. Eng. J.* **396** 125207

[218] Ding R, Qi L and Wang H 2013 Porous $NiCo_2O_4$ as an anode material for 4.5 V hybrid Li-ion capacitors *RSC Adv.* **3** 12581–4

[219] Karthikeyan K, Aravindan V, Lee S B, Jang I C, Lim H H, Park G J, Yoshio M and Lee Y S 2010 Electrochemical performance of carbon-coated lithium manganese silicate for asymmetric hybrid supercapacitors *J. Power Sources* **195** 3761–4

[220] Li F F, He Z H, Gao J F and Kong L B 2020 The investigations of pyrophosphate $CoNiP_2O_7$ produced by hydrothermal process: a high-performance anode electrode material for Li-ion hybrid capacitor *Ionics* **26** 2989–3001

[221] Cook J B, Kim H S, Lin T C, Lai C H, Dunn B and Tolbert S H 2017 Pseudocapacitive charge storage in thick composite MoS_2 nanocrystal-based electrodes *Adv. Energy Mater.* **7** 1601283

[222] Wang R, Wang S, Jin D, Zhang Y, Cai Y, Ma J and Zhang L 2017 Engineering layer structure of MoS_2-graphene composites with robust and fast lithium storage for high-performance Li-ion capacitors *Energy Storage Mater.* **9** 195–205

[223] Wang Y K, Zhang W B, Zhao Y, Li K and Kong L B 2018 Coprecipitation reaction system synthesis and lithium-ion capacitor energy storage application of the porous structural bimetallic sulfide $CoMoS_4$ nanoparticles *ACS Omega* **3** 8803–12

[224] Jagadale A, Zhou X, Blaisdell D and Yang S 2018 Carbon nanofibers (CNFs) supported cobalt-nickel sulfide ($CoNi_2S_4$) nanoparticles hybrid anode for high performance lithium ion capacitor *Sci. Rep.* **8** 1–11

[225] Chaturvedi A, Hu P, Aravindan V, Kloc C and Madhavi S 2017 Unveiling two-dimensional TiS_2 as an insertion host for the construction of high energy Li-ion capacitors *J. Mater. Chem.* A **5** 9177–81

[226] Wang Y K, Liu M C, Cao J, Zhang H J, Kong L B, Trudgeon D P, Li X and Walsh F C 2019 3D hierarchically structured CoS nanosheets: Li+ storage mechanism and application of the high-performance lithium-ion capacitors *ACS Appl. Mater. Interfaces* **12** 3709–18

Milton Keynes UK
Ingram Content Group UK Ltd.
UKHW021508140324
439454UK00003B/3

9 780750 331012